THE POWER OF PLACE
WORLD REGIONAL GEOGRAPHY

STUDY GUIDE

GIL LATZ
WITH SACHA GILBERT
PORTLAND STATE UNIVERSITY

THE POWER OF PLACE: World Regional Geography is produced by Cambridge Studios in collaboration with Australian Broadcasting Corporation, Open Learning Australia; Centre National de Documentation Pédagogìque, France; NHK School Broadcasts Division, Japan; the Teleac Foundation, the Netherlands; and Swedish Educational Broadcasting Company.

ISBN 0-471-12841-4

This book was developed for use with THE POWER OF PLACE: World Regional Geography, and Annenberg/CPB Project telecourse. This telecourse consists of 26 half-hour Video Programs broadcast on PBS; the Wiley text book *Geography: Realms, Regions, and Concepts,* written by H.J. de Blij and Peter O. Muller; this Study Guide and a Faculty Manual.

THE POWER OF PLACE: World Regional Geography is produced by Cambridge Studios in collaboration with Australian Broadcasting Corporation, Open Learning Australia; Centre National de Documentation Pédagogìque, France; NHK School Broadcasts Division, Japan; The Teleac Foundation, the Netherlands; and Swedish Educational Broadcasting Company.

Arrangements for Use
Colleges, universities, and other educational institutions may:

• Purchase the Programs on videocassette for use in the classroom or media center (1/2" VHS cassettes: $389 for the series, $39.95 per Program). For information call 1-800-LEARNER.

• Duplicate the Programs for use in the classroom or media center. For information call 1-800-LEARNER.

• License the use of THE POWER OF PLACE as a telecourse for distant learners or acquire an off-air or off-satellite taping license through the PBS Adult Learning Service. For information call 1-800-257-2578.

The Annenberg/CPB Project
The Annenberg/CPB Project was created in 1981 with the goal of increasing opportunities for Americans to acquire a high-quality college education using telecommunications and information technologies. The project leads a national movement to make innovative technology-based educational materials widely available for educational institutions, libraries, and individuals. To learn more about the project, visit our Web site at HTTP://www.learner.org.

ACKNOWLEDGMENTS

International Advisory Board

from the UNITED STATES:

H.J. de Blij, Ph.D., Professor of Geography
University of South Florida, St. Petersburg, Florida

Peter O. Muller, Ph.D., Professor of Geography
University of Miami, Coral Gables, Florida

Gil Latz, Ph.D., Professor of Geography and International Studies
Portland State University, Portland, Oregon

Osa Brand, Ph.D., Director of Educational Affairs
Association of American Geographers, Washington, D.C.

Edward Fernald, Ph.D., Professor of Geography and Associate Vice President for Research
Florida State University, Tallahassee, Florida

Paul D. McDermott, Professor of Geography and Cartography
Montgomery College, Rockville, Maryland

Richard Williams, Ph.D., Research Geologist
U.S. Geological Survey, Woods Hole, Massachusetts

from AUSTRALIA:

Mal Logan, Ph.D., Emeritus Professor of Geography and Vice Chancellor
Monash University, Melbourne, Australia

Maurie Daly. Ph.D., Professor of Geography
University of Sydney, Sydney, Australia

Kevin O'Connor, Ph.D., Associate Professor of Geography
Monash University, Melbourne, Australia

from the NETHERLANDS:

Rob van der Vaart, Ph.D., Professor of Geographical Sciences
Utrecht University, Utrecht, the Netherlands

from JAPAN:

Kenzo Fujiwara, Ph.D., Emeritus Professor of Regional Geography
Hiroshima University, Hiroshima, Japan

Yoshimi Komoguchi, Ph.D., Professor of Geography
Komazawa University, Tokyo, Japan

Mitsuru Sano, Ph.D., Associate Professor of Geography
Nihon University, Tokyo, Japan

Junji Nagata, Ph.D., Lecturer in Human Geography
University of Tokyo, Tokyo, Japan

from FRANCE:

Jacques Levy, Ph.D., Professor of Geography
Reims University and Institut d'Etudes Politiques, Paris, France

Roland Pourtier, Ph.D., Professor and Director of the Training and Research Center in Geography
Sorbonne University, Paris, France

from SWEDEN:

Solveig Martensson, Ph.D., Associate Professor of Social and Economic Geography
University of Lund, Lund, Sweden

International Co-Production Team

from the UNITED STATES, Cambridge Studios, Boston, Massachusetts

Lance Wisniewski, Co-Executive Producer
Bob Burns, Co-Executive Producer
Nancy Caulfield, Associate Project Director

from the UNITED STATES: The Annenberg/CPB Projects, Washington, D.C.
Hilda Moskowitz Goodman, Ph.D., Senior Project Officer

from AUSTRALIA: Australian Broadcasting Corporation TV and the Open Learning Agency of Australia

Peter Baroutis, Executive Producer
Tony Watts, Producer
Hardy Stow, Producer
Don Perlgut, Project Director

from the NETHERLANDS: the Teleac Educational Foundation

Anna Sepers, Project Manager
Joop van Reede, Producer/Director

from JAPAN: NHK Television, Schools Broadcast Division

Masami Yokota, Senior Producer
Kenji Tozaki, Senior Programming Director
Hideaki Nakama, Director

from FRANCE: Centre National de Documentation Pedagogique, Paris

Colette Weibel, Project Director
Pierre Carpentier, Producer
Jean-Louis Cros, Producer

from SWEDEN: Utbildningsradion, Swedish Educational Television

Ingmar Ottosson, Project Manager
Roger Samsioe, Producer/Director

THE POWER OF PLACE is produced by Cambridge Studios in collaboration with Australian Broadcasting Corporation, Open Learning Agency of Australia; Centre National de Documentation Pedagogique, France; NHK, School Broadcasts Division, Japan; the Teleac Foundation, the Netherlands; and Swedish Educational Broadcasting Company.

Major funding for THE POWER OF PLACE: World Regional Geography is provided by the Annenberg/CPB Project. The Annenberg/CPB Project was created in 1981 with the goal of increasing opportunities for Americans to acquire a high quality college education using telecommunications and information technologies. The Project leads a national movement to make innovative technology-based educational materials widely available for educational institutions, libraries, and individuals.

Additional funding for THE POWER OF PLACE has been provided by the collaborative producers listed above and John Wiley & Sons, Inc. Research funding was generously provided by the Hoso Bunka Kikin (National Broadcast Foundation), Tokyo, Japan.

The authors gratefully acknowledge the special assistance of Hilda Moskowitz Goodman, Don Perlgut, Anna Sepers, Rob van der Vaart, Masami Yokota, Francine Banner, and the faculty, students, and staff of the Portland State University Department of Geography.

Telecourse Preface

LESSONS IN GEOGRAPHY

At the end of *The Power of Place: World Regional Geography* you should be able to engage in creative and insightful discussion about these questions:

1. How and why are phenomena arranged in particular ways on the earth's surface?

2. What are the factors that create or change regions, places, environments, and landscapes?

3. How are certain events and processes shaped by human/environment interactions?

4. Describe the physical and human patterns found in the Case Study sites.

5. How is each Case Study site distinguished by a link between the human and physical world?

COURSE OVERVIEW

The telecourse you are about to view, *The Power of Place: World Regional Geography,* is the culmination of an unusual collaboration by an international team of educational broadcasters and geographers from the United States, Australia, France, Japan, the Netherlands, and Sweden. Evolving due to each country's unprecedented concern that its educational system promote greater understanding about the earth's environments and its peoples, this international video co-production has succeeded in providing a global perspective on the subject of world regional geography.

The discipline of geography is uniquely suited to address the world's human and environmental challenges because it is a field of study distinguished by a focus on the location and distribution of human and physical activity across the face of the earth, the interaction of people and environment in particular locations, and the distinctive places and regions that result. One of the striking features of geography is that little distinction is made between domestic and international matters. Such categorization of the field, in fact, has an odd ring to the professional geographer. By embracing a field of study whose goal, since the age of the Greeks has been to "write about the earth," the geographer accepts the challenge of understanding and explaining myriad forces that give shape to the world's diverse physical and social environments.

The Power of Place: World Regional Geography embraces this challenge. The telecourse provides a creative and insightful examination of the geographic forces currently shaping the complex features of the world's civilizations and environments. The telecourse includes 26 half-hour video programs — organized into 12 Units — providing penetrating insights into a range of geographic issues around the globe. All told, the telecourse visits 36 countries and over 50 sites around the world. The telecourse commences with two half-hour Introductory Programs, presenting an overview of the telecourse and highlighting the geographic concepts around which the project is organized. The remaining 24 programs that comprise *The Power of Place* contain two 10-minute documentary style case studies; each is linked together by commentary by Professor H. J. de Blij, the project's senior academic advisor and the co-author of *Geography: Realms, Regions, and Concepts: 7th Edition*, 1994 (with Peter O. Muller), the recommended companion text for the telecourse.

COURSE OBJECTIVES

To encourage the study of geography, we have used television to do what it does best: to tell compelling stories and to give you a feel for a place, something difficult to reproduce in other media. These are stories that really teach geography. Each case study tells the story of people whose lives are shaped by the geographical forces in question. Often, there are stories about research geographers trying to understand spatial variation in an era of rapid global change. In every case, we explore a vital regional and conceptual issue that can be further illuminated through geographic analysis. Through state-of-the-art computer-generated maps and animation, the programs present these issues in a highly visual and intellectually rigorous manner. Each program, then, is designed to fulfill two major goals: to characterize selected examples of regions within the 11 geographic realms explored and to communicate one or more important concepts from a variety of systematic approaches, for example, the fields of physical, political, historical, economic, and cultural geography.

TELECOURSE STUDY RESOURCES

1. The 26 half-hour Video Programs.
2. *The Power of Place: World Regional Geography Study Guide*, 1996, by Latz and Gilbert.
3. *Geography: Realms, Regions and Concepts: 7th Edition*, 1994, by de Blij and Muller.

THE TELECOURSE UNITS

Unit 1:   Introduction
Programs 1&2: Geography: A Spatial Perspective

> The Introductory Programs explore the rich variety of the world's human and physical landscapes through Case Studies focusing on observation of the earth from a NASA space shuttle and an analysis of immigration issues that confound the U.S./Mexico borderlands. The following underlying concepts of the telecourse are introduced:
> - **Relative Location**
> - **Human/Environment Interaction**
> - **Realms and Regions**
> - **Scale**
> - **Spatial Perspective**

Unit 2:   Europe: Confronting New Challenges
Program 3:   Supranationalism and Devolution

Strasbourg: The coexistence of French and German culture in France's eastern borderland.

Slovakia: The recent birth of Slovakia and the Czech Republic.

Program 4:   **East Looks West**
Berlin: The transition from a weakened and divided city to one of
                          emerging importance.
Poland: The diffusion of democratic ideas throughout Poland.
Program 5:   **The Transforming Industrial Coreland**
Liverpool: An analysis of this port city's once-thriving industrial
                          economy, fallen to its marginal present-day role.
Randstad: The emergence of this region in the Netherlands as an
                          integral part of the current European core.
Program 6:   **Challenges on the Periphery**
Iceland: The many challenges faced by Iceland, which exists on the
                          cultural and physical periphery of Europe.
Andalucía: The gradual decline of the once central but now peripheral
                          Andalucía region of Spain and its hopes for the future.

Unit 3:   **Russia's Fracturing Federation**
Program 7:   **Facing Ethnic and Environmental Diversity**
Dagestan: The question of independence in the ethnically mixed
                          Republic of Dagestan in the face of devolutionary processes.
Vologda: The uncertainties of a harsh climate and poor infrastructure
                          in the Russian countryside around Vologda.
Program 8:   **Central and Remote Economic Development**
St. Petersburg: The effects of the shift to a market economy on real estate
                          values in St. Petersburg.
Bratsk: The difficulties of industrial production in Siberia.

Unit 4:   **North America: The Post-Industrial Transformation**
Program 9:   **Inner vs. "Edge" Cities**
Boston: The circumstances that prompt formulation of an
                          Empowerment Zone proposal in inner-city Boston.
Suburban Chicago: The pressure of suburban growth on agricultural
                          communities surrounding Chicago.
Program 10:   **Ethnic Fragmentation in Canada**
Quebec: The resistance of French speakers to domination by the
                          English language in Montreal, Quebec.
Vancouver: The influence of immigration by Hong Kong Chinese on
                          this emerging Pacific Rim metropolis.
Program 11:   **Regions and Economies**
Oregon: The competition for water resources in eastern Oregon.
Midwest Auto: The incorporation of Japanese production techniques in the
                          midwest U.S. automotive industry.

Unit 5: The Geographic Dynamics of the Western Pacific Rim

Program 12: The Japanese Paradox: Small Farms and Mega-Cities

Northeast Japan: The natural hazards that can affect the process of irrigated rice production are explored.

Tokyo: The mass transportation characteristics of Japan are portrayed through a profile of the commuting regimen of a Tokyo businessman.

Program 13: Global Interaction

Singapore: A study of how Singapore exploits its location to play a key commercial role in Pacific Asia.

Australia: An exploration of Australia's European roots and recent Asian influences on economic development and trade.

Unit 6: Middle America: Collision of Cultures

Program 14: Migration and Conquest

Mexico: A review of migration patterns and causes both within and outside Mexico.

Guatemala: A portrait of the "cycles of conquest" borne by Mayan peoples in Guatemala.

Unit 7: South America: Continent of Contrast

Program 15: Andes and Amazon

Ecuador: Efforts to monitor and ameliorate the negative effects of volcanic activity in Ecuador.

Amazon: A review of the concept of sustainable development for the Amazon in Northeast Brazil's Parà State.

Program 16: Accelerating Growth

São Paulo: The dilemmas facing urban homesteaders in São Paulo, Brazil.

Santiago: The lure of an export economy for Santiago, Chile.

Unit 8: North Africa/ Southwest Asia: The Challenge of Islam

Program 17: Sacred Space Under Siege?

Jerusalem: The application of a spatial perspective to explain religious conflicts in Jerusalem.

Istanbul: Gaps between rich and poor, secular and fundamentalist, are explored in Istanbul, Turkey.

Program 18: Population, Food Supply, and Energy Development

Egypt: Rapid population growth in Cairo challenges agricultural productivity and food supply in Egypt.

Oman: Oil revenue, Muslim life, and the diversification of the economy are addressed as Oman attempts to modernize itself.

Unit 9:   Subsaharan Africa: Realm of Reversals

Program 19:   The Legacy of Colonization

Ivory Coast: The problems resulting from over reliance on cacao production are explored in Ivory Coast.

Gabon: Challenges accompanying energy development in Gabon, and the limitations imposed by an inadequate infrastructure.

Program 20:   Understanding Sickness, Overcoming Prejudice

Kenya: A spatial look at healthcare and disease in Kenya.

South Africa: New land reform policies in post-Apartheid South Africa strive to address social and economic inequities.

Unit 10:   South Asia: Aspiring India

Program 21:   Urban and Rural Contrasts

Delhi: A geographer studies Delhi as a multicultural, rapidly growing metropolitan area.

Dikhatpura: Local farmers' lives are transformed through the provision of irrigation in a water-deficient area.

Unit 11:   China and Its Sphere

Program 22:   Life in China's Frontier Cities

Lanzhou: Both China's Muslim minority and the environmental character of its Northwest region are profiled in Lanzhou, China.

Shenyang: The development challenges facing China's traditional industrial heartland are discussed in Shenyang, China.

Program 23:   China's Metropolitan Heartland

Shanghai: Physical location at the mouth of the Chang Jiang River and new government policies bolster development in Shanghai.

Nanjing: Foreign investment establishes a garment factory in an agricultural village outside Nanjing, creating new employment opportunities.

Program 24:   The Booming Maritime Edge

Guangdong: A review of Nike's use of the Global Production System in Guangdong Province.

Taiwan: This newly industrializing economy charts its future through the creation of sophisticated science parks producing high-technology products.

Unit 12:   Southeast Asia: Between the Giants

Program 25:   Mainland Southeast Asia

Laos: Prospects for development arise in isolated Laos as a result of a new bridge spanning the Mekong River.

Vietnam: Government policies provide incentives for increasing rice production in the Mekong delta.

Program 26:   Maritime Southeast Asia

Indonesia: The growing importance of tourism confronts the society and economy of Indonesia.

Malaysia: Policies balancing ethnic diversity are defined as the key to Malaysia's recent economic success.

THE POWER OF PLACE STUDY GUIDE

The purpose of the Study Guide is to aid students in their study of the material presented in *The Power of Place*. Each Unit of this course explores a specific realm of the world and analyzes it through one or more 30-minute programs composed of two cases linked together with interview commentary by Professor H.J. de Blij that highlights several of the themes and concepts under review. The Study Guide structure for program-by-program examination of geographical themes and concepts is in two major sections as follows:

Before You View the Video Program

LESSONS IN GEOGRAPHY

Lists the main learning objectives for each Program.

OVERVIEW

Provides summaries of each Case Study to be shown.

STUDY RESOURCES

Identifies the written and video material covered for each program.

PREVIEWING QUESTIONS

Clarifies some of the terminology and themes with which you should become familiar prior to viewing each Video Program.

KEY THEMES

Identifies the key themes that comprise the visual and narrative structure found in each Program's Case Studies.

DISCUSSION OF CASE STUDY THEMES

Discusses the key themes that comprise the visual and narrative structure found in each Program's Case Studies.

GEOGRAPHERS AT WORK

Profiles the application of geographic knowledge by several of the professional geographers associated with the telecourse.

After You View the Video Program

CASE STUDY CONNECTIONS

Compares key points of the Program raised in each Case Study.

TEST YOUR UNDERSTANDING

Asks questions regarding the issues discussed in each video, and identifies map exercises that pinpoint the geographic characteristics of each program.

MAP EXERCISES

The student should note that, since this is a world regional geography course, map review and map exercises are indispensable to absorption of the course material. The **Test Your Understanding** section requires that you use a variety of cartographic materials to explore the characteristics and relationships that distinguish the places portrayed in the telecourse. Two sets of maps are contained in this publication — Case Study Overview maps and Base maps. These maps are provided at the beginning of each Unit and can be used to complete the **Map Exercises**. You may find extra copies of these maps helpful. In addition, the student and instructor are urged to make connections between the case study themes and the thematic maps found in the recommended companion text. An atlas of the world will also facilitate student learning of the ideas presented in *The Power of Place*.

CONSIDER THE DISCIPLINE

Links the telecourse to the companion text, *Geography: Realms, Regions, and Concepts: 7th Edition*, 1994, by de Blij and Muller, through incorporation of systematic analyses of conceptual issues in world regional geography.

CRITICAL VIEWING

Suggests written essay topics based on the Case Studies in each Program.

WHAT IS A TELEVISION COURSE

The primary difference between television courses and traditionally taught courses is the manner in which instruction is delivered to students. Traditional students may come to campus for direct instruction by a faculty member several times per week; television course students work more independently, watching the television programs and reading the print materials at home or at work with guidance from the course faculty through a variety of communication and instructional techniques. The number of required class meetings, however, is generally much fewer than in traditionally taught classes because the basic course content is delivered to students through television, print, and other types of faculty/student interactions.

Television courses, in most other respects, e.g., academic rigor, student requirements, and need for qualified faculty, are equivalent to traditionally taught college courses. Television course students enroll in the college or university that has adopted the course, pay tuition to that institution, usually have access to student services offered at that institution, receive academic credit from that institution and are taught by that institution's faculty. These faculty members make similar kinds of academic and instructional decisions as they would in traditionally taught classes and interact with students through a combination of classes, written assignments, phone contacts, and mailings.

TAKING THE TELECOURSE

Find out the following information when you register:

- What books are required for the course?

- Are orientation sessions scheduled?

- When will *The Power of Place* be broadcast in your region?

- When are course examinations scheduled?

- Are any on-campus meetings scheduled?

Be an active learner:

1. Before viewing each Video Program, read the sections of your Study Guide titled **Lessons in Geography, Overview, Previewing Questions**, and **Discussion of Case Study Themes**.

2. Read the textbook assignment listed in the Study Guide and any supplementary material assigned by your instructor. Note any overviews, lists of key ideas and concepts, and summaries which will help you identify important information.

3. View the program, keeping the **Lessons in Geography** in mind. Be an active viewer. You may want to record the program on a VCR if you own one. If you don't have a VCR, you can make an audiocassette for review. Some public television stations also repeat programs during the week. Some students find it helpful to take notes while viewing the program.

4. Complete the **After You View the Video Program** section and take notes on the review questions. Your instructor may ask you to submit written replies to some of the questions.

5. Review the **Critical Viewing** questions. Complete any other questions, activities, or essays assigned by your instructor.

6. Keep up with the course on a weekly basis. Each Unit of the course builds on knowledge gained in previous units. Stay current with the programs and readings. Make a daily checklist and keep weekly and term calendars, noting scheduled activities such as meetings or examinations, and set aside blocks of time for viewing programs, reading, and completing assignments.

7. Keep in touch with your instructor. If possible, get to know him or her. You should have your instructor's mailing address, phone number, and call-in or office hours. Your instructor would like to hear from you and to know how you are doing, and he or she will be eager to answer any questions you have about the telecourse.

Contents

Unit 1   **Introduction**   1

Programs 1 & 2   Geography: A Spatial Perspective   5

Unit 2   **Europe: Confronting New Challenges**   13

Program 3:   Supranationalism and Devolution   19

Program 4:   East Looks West   25

Program 5:   The Transforming Industrial Coreland   33

Program 6:   Challenges on the Periphery   39

Unit 3   **Russia's Fracturing Federation**   45

Program 7:   Facing Ethnic and Environmental Diversity   51

Program 8:   Central and Remote Economic Development   57

Unit 4   **North America: The Post-Industrial Transformation**   63

Program 9:   Inner vs. "Edge" Cities   69

Program 10:   Ethnic Fragmentation in Canada   77

Program 11:   Regions and Economies   83

Unit 5   **The Geographic Dynamics of the Western Pacific Rim** 93

Program 12:   The Japanese Paradox: Small Farms and Mega-Cities99

Program 13:   Global Interaction   107

Unit 6   **Middle America: Collision of Cultures**   115

Program 14:   Migration and Conquest   121

Unit 7   **South America: Continent of Contrast**   127

Program 15:   Andes and Amazon   133

Program 16:   Accelerating Growth   139

Unit 8   North Africa/Southwest Asia: The Challenge of Islam 145

Program 17:   Sacred Space Under Siege?   151

Program 18:   Population, Food Supply, and Energy Development 157

Unit 9   Subsaharan Africa: Realm of Reversals   163

Program 19:   The Legacy of Colonization   169

Program 20:   Understanding Sickness, Overcoming Prejudice   175

Unit 10   South Asia: Aspiring India   181

Program 21:   Urban and Rural Contrasts   187

Unit 11   China and Its Sphere   195

Program 22:   Life in China's Frontier Cities   199

Program 23:   China's Metropolitan Heartland   205

Program 24:   The Booming Maritime Edge   211

Unit 12   Southeast Asia: Between the Giants   217

Program 25:   Mainland Southeast Asia   223

Program 26:   Maritime Southeast Asia   229

▪ UNIT 1
Introduction

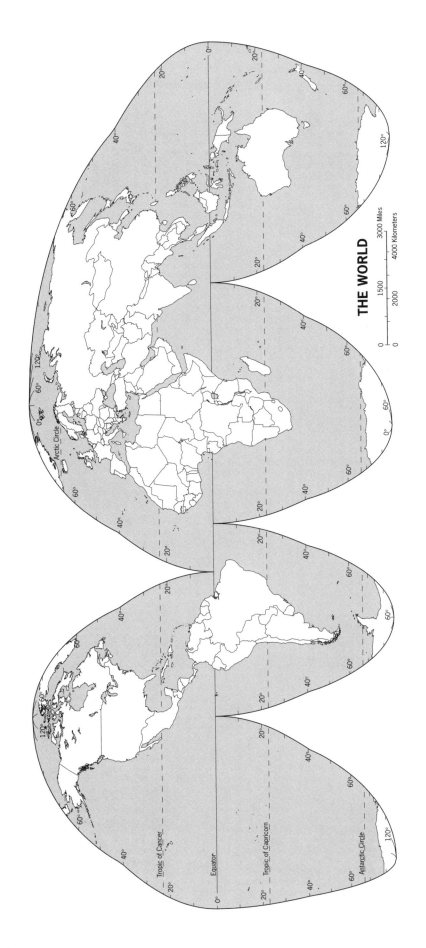

THE WORLD

0	1500	3000 Miles	
0	2000	4000 Kilometers	

Arctic Circle

Tropic of Cancer

Equator

Tropic of Capricorn

Antarctic Circle

▪ PROGRAMS 1 & 2
Geography: A Spatial Perspective

LESSONS IN GEOGRAPHY

At the end of Introductory Programs 1 & 2, you should be able to:

1. Understand the structure of *The Power of Place*: The 12 Units cover the 11 realms of world regional geography and their 26 Programs visit over 50 places around the globe.

2. Explain how geographers look at the world.

3. Describe the use of changing scale — the geographer's tool — as a device to better understand issues of unity and division.

4. Highlight the general distribution of population on the earth's surface.

5. List the five concepts of geography required for this telecourse.

6. Apply a spatial perspective to the borderland issues presented in Program 2.

OVERVIEW

The Power of Place: World Regional Geography has two main objectives. First, the course introduces the field of geography, a discipline that links human societies to their natural environments. Second, it investigates the great **geographic realms** of the modern world and examines their human and physical contents, assets and liabilities, connections and barriers, and potential and prospects for the future.

The telecourse examines 11 major realms of the world and explores their interconnections. What are the physical and human patterns that can be found? How and why are myriad phenomena — people, vegetation, climates, mountains, cities — arranged in particular ways on the earth's surface? What factors are involved in creating or changing particular regions, places, environments, or landscapes? Perspectives from physical, political, historical, economic, and cultural geography are used to characterize regions that comprise each realm. In this way, viewers gain insight into how the world's regions are knit together into a spatial framework.

The introductory programs raise two questions: 1) What insight can geography provide regarding the unity and division that pervades the contemporary world? and 2) What are the concepts of geography that help us analyze characteristics of the world's regions?

Maps are analytical tools unique to geography are used extensively throughout the telecourse. The cartography used in Program 1 illuminates the grand diversity of our world and highlights patterns of physical and human activity across the earth's surface. Maps are essential to any discussion of contemporary world affairs from a geographical perspective. As Professor H.J. de Blij notes, geographers look at the world spatially. They pose their problems spatially. They seek their solutions spatially, and for this they use maps. Maps tell us how the world is laid out.

Throughout these introductory programs we see glimpses of later Case Studies. We travel to Lanzhou at the threshold of the rugged and isolated Northwest region of China to experience the contrasting lives and livelihoods of its Muslim and Han Chinese inhabitants and to Iceland, where we discover the ways that climate and ocean currents in a particular location have influenced the prosperity and recent decline of the local fishing industry. We journey to Jerusalem, to examine how a distinctive religious landscape is at the heart of Israeli/Palestinian territorial conflict and to Tokyo, to observe how mass transportation serves the needs of one world mega-city. Near Chicago, we learn how competition between urban and agricultural interests at the city's edge threaten prime agricultural land and in the small farming village of Dikhatpura, India, we see the impact irrigation has had on agriculture and living conditions.

Program 1 asks the provocative question: Is the world now both more divided and more united than ever before? Today, unity is symbolized by the fall of the Berlin Wall, the most famous icon of the Cold War rivalry between the Free World and the former Communist Bloc. Growing division can be seen in state **devolution** as separatist movements bubble up in such diverse areas as Russia, the Balkans, Quebec, and even the western United States. One answer which balances unity and division is offered by geographer David Ley in our visit to British Columbia, Canada, seen in full in Program 10. Here cultures clash over the differing housing style preferences of Anglo residents and new Chinese immigrants. Ley points out that this case is a striking example of the global meeting the local in a Vancouver residential neighborhood. This meeting of global and local is proposed as the reason why the world is both more united and more divided than ever before: local conflict, devolution, and separatism are now magnified by globalization.

Program 2 introduces the telecourse's Case Study approach by studying the borderland region between the North American and Middle American realms. A journey to Anapara and Campestre near Cuidad Juarez, Mexico and then across the Rio Grande to El Paso, Texas walks us through the five concepts of geography which will be used throughout this telecourse.

STUDY RESOURCES

1. Video Programs 1 & 2: "Geography: A Spatial Perspective" from *The Power of Place: World Regional Geography*.
2. Introduction "World Regional Geography: Physical and Human Foundations" from de Blij and Muller, 1994, *Geography: Realms, Regions, and Concepts: 7th Edition*.

PREVIEWING QUESTIONS

1. Do you have access to a World Regional Geography textbook and an atlas?
2. Describe the layout of the world's major geographic realms.
2. Why are maps crucial to the study of geography?
3. Define unity, supranationalism and devolution.
4. What does the term globalization mean?

TELECOURSE CONCEPTS

- **Relative Location**
- **Human/Environment Interaction**
- **Realms and Regions**
- **Scale**
- **Spatial Perspective**

DISCUSSION OF TELECOURSE CONCEPTS

• Relative Location

Location is one of the central concepts in geography. Geographers typically describe places in terms of their absolute and relative location. The absolute location of a place can be defined in terms of its latitude and longitude, that is, a mathematically established network of lines projected onto the earth's surface. In simplest terms, this system of measuring allows us to give an unique address to every place on the earth.

On the other hand, relative location exhibits functionality. This is an extremely powerful idea which implies that location is a function of spatial interaction, interdependence, and accessibility. The location of one place relative to another tells us much about forces that lead to urban settlement and urban growth.

Program 2 focuses on the relationship between the cities of Ciudad Juarez, Mexico, and El Paso, Texas. How Juarez's location is relative to El Paso and the U.S. border answers a functional question: Why have *maquiladora* assembly plants been located in this particular place? The answer can be found in the relative location of a growing, job-seeking population in one area which supplies inexpensive labor for goods destined for another locale.

As Professor de Blij notes in his commentary, one of the most useful ways to assess the fate or fortune of a place is to look at its relative location; where it is vis-à-vis other places. And, as the film shows, the concept of relative location helps us understand a whole set of patterns on the surface of the earth, ranging from the distribution of parts suppliers for the U.S. auto industry, to the perceived strategic location of St. Petersburg, and the site of Lanzhou along the historic Silk Road.

• Human/Environment Interaction

The interaction of humans and the environment, or the relationship between human society and the natural environment has two dimensions: how people adapt to their local environment, and how people overcome the limitations of that environment. Emphasis on the idea of interaction is essential to the geographical perspective, for the field of geography is just as interested in the physical characteristics of places as it is in inhabitants' adaptation to and exploitation of those places.

Geographers have long recognized that environmental circumstances influence the distribution of people across the face of the earth. Despite technological achievements, the distribution of the world's population is heavily influenced by precipitation, temperature, and soil type. Space shuttle astronaut Mike Foale observes that, from space, vast areas of the world seem to be one color: brown. Foale says that the most striking thing one notices about the world from outer space is that places to live in the world are few and far between, and we are already living in all of them.

Resource endowments and the knowledge required to manage those endowments is not equally distributed throughout the world. Though Lanzhou, a Case Study site discussed in depth in Program 22, is understandable in terms of its location relative to the Silk Road; it is equally important to observe that the people living in this semi-arid environment have cleverly adapted to their surroundings through careful management of rich, though highly fragile and erodable, löess soils. Assessment and adaptation to the local conditions is an example of the concept of human/environment interaction.

• Realms and Regions

Realms are the broadest units of geographic classification. In *The Power of Place* the world is divided into 11 realms: North America, Middle America, South America, North Africa/Southwest Asia, Subsaharan Africa, South Asia, China, Southeast Asia, the Pacific Rim of Austrasia, Russia, and Europe.

Within realms we find regions, areas of realms marked by a certain degree of homogeneity. Regions are parts of the earth's surface which share some common feature or trait, for example, the spruce forest zone that characterizes the vegetation of the Pacific Northwest portion of North America.

The regionalization of a realm is a complex exercise, as illustrated by our visit to Ivory Coast in the region of West Africa. This area is distinguished by primarily French and secondarily British colonial activity, coastal-oriented states, large total populations, and a distinct set of east-west ecological belts that serve as a backdrop for the location of the cacao plantations and the migration of various ethnic peoples from the interiors of Western African countries to take advantage of employment in the better endowed environmental zones of Ivory Coast. Colonial heritage, rapid population growth, and environmental abuse also underscore the regional identities of other regions of Africa; Eastern Africa, a region in which Kenya is located below the Sahara Desert, has not only these problems but a number of other ecological crises such as trachoma, malaria, and a rapidly growing AIDS epidemic.

The geographer's concern with regions can help to explain the characteristics and problems of development within realms, as well as the connection between regions and realms, where combinations of domestic and international factors continue to shape the livelihood of a significant portion of the world's population.

• Scale

The concept of scale may be the most complex and yet the most utilitarian of the tools used by geographers to analyze the world. For example, scale is indispensable to the creation of maps. Scale refers to the ratio of the distance between two places on a map and the actual, real world distance between those places. Maps represent varying levels of generalization with respect to the earth's surface depending on the scale used — the larger the area represented, the smaller the scale. As scale changes, so too does the level of detail about the grain of the earth's surface; large scale maps portray smaller portions of the earth's surface; small scale maps, just the opposite. Thus, large scale maps have more detail about the surface of the earth than do small scale maps.

The concept of scale is also used to describe the world encountered by geographers in the field. The information used to make generalizations in the field changes, relative to the object of observation, be it streetscape, neighborhood, city, metropolitan area, state, nation, or region. To the geographer, there is an important correlation between the extent of an area, the scale of measurement, and the detail captured. As seen in the video, if one views Australia at the scale of an individual street in Fairfield, the country's Asian population seems predominant. In contrast, when the country is viewed at the scale of the larger metropolitan area of Sydney, one sees an urban area predominately settled by Europeans. As scale changes to that of the nation as a whole, and then the Western Pacific region, the historical, economic, and cultural interaction of the Australian continent with the rest of the world takes on a new meaning derived from a complex spatial perspective.

• Spatial Perspective

The spatial dimensions of human experience is a distinguishing facet of the geographical perspective. But what does this mean? The concepts of relative location, human/environment interaction, regions and realms, and scale all combine to create a spatial perspective. To think in spatial terms, then, is to think like a geographer; to understand what is located where, as well as how and why, spatial variation can be identified across the face of the earth. The acquisition of a spatial perspective means that one has the ability to describe and analyze spatial organization — the patterns of people, places, and environments — on the earth's surface.

By means of this spatial perspective Program 2 challenges us to look at the realms and regions shared by the borderland cities of Ciudad Juarez and El Paso in new ways. Relative location is both the background for the location of Juarez and the reasons for the dispersion of Latino people and culture across the boundary between the U.S. and Mexico. The prospects for future rapid growth of the settlements on both sides of the boundary can be examined in terms of the less-than-careful water management which takes place as humans interact with this semi-arid environment. The study of this area at the meeting of the realms of North America and Middle America — one developed and the other developing — helps us to understand the motivation to migrate and showcases the unique issues facing people who live in a zone which combines the elements of two different realms. Using the concept of scale, one can organize a multiplicity of images into a complete picture of rich and poor communities coexisting in the same metropolitan region but not, of course, in the same exact location.

A spatial perspective grants insight into the reality of a borderland region. Through this perspective one can not only quite literally "see in the dark" as illegal immigrants are tracked and apprehended, but can understand broad patterns of spatial organization on the surface of the earth in new and profound ways.

GEOGRAPHERS AT WORK

Geographers are fascinated with why the world looks the way it does. Program 1 shows how maps, site-level information, and a spatial perspective provide NASA with a greater understanding of the world and its problems. Geographers use satellite observation photography and other imaging processes, as well as data collection in the field, to construct cartographic representations of the world in which we live. It is through maps that geographic analysis grants insight into a range of contemporary issues, from ethnic conflicts to environmental pollution, two phenomena which seem to be growing in intensity around the world.

Geographers use maps to provide a generalized understanding of places through description as well as to specify relationships between places. But in the effort to understand such relationships, the geographer must be concerned first and foremost with the distinctiveness of different places; this principle of geographical research makes field work indispensable to the art of generalization and policy recommendation.

Arguably the most unusual form of field work in which a geographer might engage in is space travel. The geographer at work in Program 1 does just that, not as an astronaut, but as an expert in earth observation, one of the important objectives of space shuttle research. Justin Wilkinson, a physical geographer with a doctorate from the University of Chicago, teaches astronauts something about what the world looks like and what major issues arise in the sciences today. By understanding the distinctiveness of different places, astronauts are able to conduct a unique kind of field work through photography from outer space.

When this geographer and his students look at the range of global problems, there is recognition that the challenges — and the solutions — to overcoming these problems can be found at the local level.

TEST YOUR UNDERSTANDING

Questions

1. The Rio Grande River clearly marks a part of the border between Mexico and the United States. On any map of the Western Hemisphere, identify some other borders you can see. Name borders you can't see.

2. How do people shape places? How to places shape people?

3. Give three examples from the video of how changing the scale of observation can change your perception of a place.

4. Explain how landscapes vary among realms. Explain how landscapes vary within realms.

5. Why do astronauts learn about geography?

Map Exercises

Refer to any standard atlas or the maps in the Introduction of "World Regional Geography: Physical and Human Foundations," from de Blij and Muller, 1994, *Geography: Realms, Regions, and Concepts: 7th Edition*, to assist in this exercise.

1. Place the following names:

Rio Grande River	Cuidad Juarez	El Paso
Lanzhou	Ivory Coast	Tokyo
Jerusalem	Chicago	Vancouver, B.C.
St. Petersburg	India	Kenya
Sydney	Quebec	Berlin

CONSIDER THE DISCIPLINE

REGIONAL AND SYSTEMATIC APPROACH DIVIDES AND UNITES GEOGRAPHY

Geographers often remark that their field of study is, literally, everywhere. It can be seen from any window. But the unanimity with which geographers embrace the heritage of the discipline as the science that strives "to write about the earth," also masks fundamental divisions of opinion about the methodologies for doing so. These differences of opinion are captured in debates about whether geography should be defined by its **regional** or **systematic traditions**.

Regional geography is considered as a key unifying paradigm of the field, around which all other systematic approaches, for example, urban, economic, or political, are organized. Systematic geographers counter, however, with persuasive arguments that regional geographic thought is often only descriptive, incapable of creating scientifically rigorous **models** for testing the **hypotheses** that are presumed to be at the heart of all the social scientific disciplines.

This problem of **differing methodological approaches** is mediated in the de Blij and Muller companion text to this telecourse through incorporation of systematic analyses in each unit covering a world realm. This approach allows for the broadest possible discussion of the discipline of geography and highlights one of the most distinctive qualities of the field. Through both regional and systematic geography, the discipline is equally curious about and capable of contributing to **physical scientific investigations** as well as **social scientific investigations,** increasingly in the interesting and important world of **ecological studies** which straddles the physical and social dimensions of the world in which we live.

For further discussion of regional and systematic aspects of the field of geography, as represented by such topic areas as geomorphology, biogeography, and cultural, political, and economic geography, see the section on Regional and Systematic Geography and their Relationships, and What do Geographers Do?, in the Introduction from de Blij and Muller, 1994, *Geography: Realms, Regions, and Concepts: 7th Edition.*

CRITICAL VIEWING

Based on the Case Studies in Video Programs 1 & 2 "Geography: A Spatial Perspective" from *The Power of Place: World Regional Geography,* and the reading in Introduction, "World Regional Geography: Physical and Human Foundations," from de Blij and Muller, 1994, *Geography: Realms, Regions, and Concepts: 7th Edition,* develop essays answering the following questions:

1. Based on current world events which you have been following in the media, describe how local conflict is magnified by globalization.

2. Explain how global issues, such as ozone depletion or international trade, ignore national boundaries.

3. As noted by astronaut Mike Foale, people are not distributed evenly across the earth. Explain how this uneven population distribution leads to environmental degradation.

▪ UNIT 2

Europe:
Confronting New Challenges

RESILIENT EUROPE: Confronting New Challenges

| Kilometers | 0 | 500 | 1000 |
| Miles | 0 | 250 | 500 |

▪ PROGRAM 3

Supranationalism and Devolution

LESSONS IN GEOGRAPHY

By the end of Program 3, you should be able to:

1. Recognize the border-area interaction between Strasbourg, France and the Alsace area of France and Germany.
2. Determine the part Strasbourg plays in the process of European unification.
3. Understand that pivotal role Strasbourg plays in the "Blue Banana" zone of Western Europe.
4. Briefly review the history of Slovakia and its separation from the Czech Republic.
5. Identify the centrifugal forces emerging in Slovakia.

OVERVIEW

Strasbourg: Europe's New Capital? focuses on the **coexistence** of French and German **cultures** in the city of Strasbourg. This peaceful and prospering community remains an important symbol of the new **European unity**.

The city has been influenced primarily by French and German cultures, and its landscape and architecture have recorded the periodic dominance of one or the other of these cultures. Today, people from France and Germany are crossing the Rhine River without border controls, working in industries and services throughout the border region, and may even own land in both countries. These freedoms have not always existed, as seen in interviews with local residents describing pre-existing border procedures, migration patterns, and land ownership. The recent transformation and relaxation of **political territorial control** in this part of Europe provides an example of the new unification of Europe.

Slovakia: New Sovereignty explores the recent birth of Slovakia and the Czech Republic and the problematic **transition** from the old Czechoslovakia to these two new states. The young political state was created in January 1993 with the breakup of the old Czechoslovakian Federation. Located northeast of Austria and north of Hungary, this area has a long frontier history on the edge of Western Europe. A study of the town of Bratislava provides a window into the frontier history shaping this new political state.

The political decision to separate remains controversial. From the perspective of an historian, the separation was a Czech political class decision. A newly independent Slovakia could provide further reinforcement of the perceived **barrier** between the new Czech Republic, the neighboring Ukraine, and the now-more-distant Balkans. This insulation would allow the Czech economy to conform more easily to those of its Western European neighbors.

While the leadership of each state supported peaceful separation, leading to the formation of two independent countries, the major question emerging was if the mandate was based upon the **popular will** of the people. Interviews with different members of both societies show a varying range of opinions. The new border affects not only the lives of local inhabitants, like those of Sedonia, but also the economic and political spheres that have long been operating in the old Czechoslovakia.

STUDY RESOURCES

1. Video Program 3: "Supranationalism and Devolution" from *The Power of Place: World Regional Geography*.
2. Unit 2 Map, "Resilient Europe: Confronting New Challenges" from Latz with Gilbert, 1996, *The Power of Place: World Regional Geography Study Guide*.
3. Chapter 1, "Resilient Europe: Confronting New Challenges," from de Blij and Muller, 1994, *Geography: Realms, Regions, and Concepts: 7th Edition*.

PREVIEWING QUESTIONS

1. Explain the role rivers, oceans, and seas play in the development of a state.
2. Define irredentism, nationalism, nation, and state.
3. Locate the Rhine and Danube Rivers.
4. Locate the states of the former Communist Bloc on a map.

Germany + France

KEY THEMES
- **Borders and Interaction**
- **Europe Under Construction – Supranationalism**
- **Nation-State Boundaries**
- **Cultural Diversity**

DISCUSSION OF CASE STUDY THEMES
- **Borders and Interaction**

Over the last century, Strasbourg has gone through five changes of nationality and three wars, two of which involved the world powers. For centuries, the frontier city has been a conquest objective of both France and Germany, and the Rhine River has always served as the interactive link between the inhabitants of the German Rhineland and adjacent French areas.

Strasbourg's central location, relative to the counties of Western Europe, has played a major role in the formation of the European Union (EU). One facet of this union is discussed through an interview with the Deputy General Secretary of the Council of Europe. He explains this forum's promotion of democracy and supranational issues in Europe. The Council also addresses the economic, cultural, and political concerns of the European Union.

• Europe Under Construction – Supranationalism

Strasbourg has always been the "town at the crossing of the roads," leading to the Atlantic Ocean from Central Europe and linking the British Isles to the Mediterranean Sea. Strasbourg is part of the "Blue Banana," a zone of development starting in London passing through Belgium, France's Alsace region, the Rhineland, and Switzerland before ending in Milan.

Strasbourg is the site of three supranational organizations: 1) The European Parliament — an elected 410-member legislature of the European Union; 2) The European Court of Human Rights — a humanitarian and legal rights institution; and 3) The Council of Europe — a forum for the promotion of political democracy throughout Europe. Near Strasbourg and among the other members of the European Union, border crossings are no longer subjugated to customs control. Strasbourg serves as a symbol of a peaceful and contiguous Europe.

While the city is vibrant economically, political changes outside the country threaten its position in the Blue Banana. Alsace represents the only French provincial area connected to the region, and it is an important component in the French economy. The opening of Eastern Europe now threatens to displace this region, as the process of development advances in the former Communist Bloc.

• Nation-State Boundaries

In 1993, the newly elected Prime Ministers of Czechoslovakia and Slovakia decided that the two states should separate peacefully for two basic reasons: so that Slovakia could aid the Czech Republic by serving as a buffer between the Republic of Ukraine and the Baltic states and because Slovakia did not agree with the decision by the Czech state to conform their economy to Western European norms. The official border between the states is in dispute, however, and a bilateral commission has been working for two years to settle the situation. There are mixed feelings among the people of Slovakia; some say the people were not consulted as to whether they supported the separation. In Sedonia, the main road is the dividing line which has displaced people's property on both sides of the border. A border checkpoint has been instituted further dividing the people of the region.

• Cultural Diversity

For the Czechs, the abolition of German domination and the liberation of Slovakia from Hungary were centripetal forces which gave birth to the common state at the end of World War I in 1919. Freed from Nazi control at the end of World War II, Czechoslovakia came under Soviet domination and the power of a communist regime. With the end of the Cold War, these forces are now obsolete, and the Czech Republic is turning itself toward Western Europe as Slovakia, due to its geographic position, naturally aligns itself with Eastern Europe.

Centrifugal forces now threaten both European Union and the new state of Slovakia. The Hungarians are accusing the Slovaks of displacing the border between the two states. In Western Europe the notions of territory, state, and nation have been clearly defined for centuries, but in central Europe they are still unclear. Slovakia also faces the threat of irredentism with respect to its Hungarian minorities living along the north bank of the Danube River.

There are about 600,000 Hungarians living in Slovakia, and people inhabiting the regions peacefully coexist. In towns, parents can chose to send their children to the ethnic school of their choice, and about two–thirds of the children are enrolled in Slovakian schools. Again, mixed feelings arise regarding cultural diversity. Hungarians and Slovaks claim to be compatible, however, both groups still discriminate against the Gypsies.

During the Communist era, the government dealt with minority problems by denying their existence, but now that the country is in the process adapting to a market economy, it is tempted to designate scapegoats to mask real problems.

CASE STUDY CONNECTIONS

1. Both Strasbourg and Slovakia grapple with political territorial issues.
2. Transnational — ethnic, economic, and political — issues dominate contemporary affairs in each place.
3. Relative location helps to explain the cultural landscape of each place.
4. Rivers act as important natural boundaries, transport arteries, and providers of energy in each place.

GEOGRAPHERS AT WORK

Between the 52 Case Studies and at the end of the 26 Programs in this series, Harm de Blij, comments on the themes seen in the videos — and sometimes, a few facts and concept that aren't.

Professor de Blij is known to millions as the Geography Editor of *Good Morning America* on ABC. Formerly a Distinguished Professor of Geography in the School of Foreign Service at Georgetown University and now on the faculty of the University of South Florida, de Blij is the senior author of several textbooks, including a companion text to this series, *Geography: Realms, Regions, and Concepts: 7th Edition.*

Early in his career as student of physical geography, he spent a year in the field in Swaziland invesitgating the geomorphology of the African Rift Valley. While preparing for this research at Northwestern University in Evanston, Illinois, classes in political geography caught his attention and never let go. His blend of physical and political geography brings a critical perspective to *The Power of Place* telecourse.

TEST YOUR UNDERSTANDING
Questions

1. How has Strasbourg's relative location affected its industrialization both positively and negatively?
2. Why have Strasbourg and Slovakia been the targets of the geo–political aggressions of dominant states?
3. Define centrifugal and centripetal forces and explain how they differ.
4. Explain how cultural diversity marks these two locations.

Map Exercises

The maps provided at the beginning of this unit can be used to complete this exercise. You may find extra copies of these maps helpful. Also, refer to the maps of Chapter 1, "Resilient Europe: Confronting New Challenges," from de Blij and Muller, 1994, *Geography: Realms, Regions, and Concepts: 7th Edition*, and any standard atlas to assist in this exercise.

1. Place the following names:

Rhine River	Danube River	Bratislava
Strasbourg	London	Milan
Ukraine	Poland	Germany
France	The Czech Republic	Austria
Hungary	Alsace	Prague
Slovakia		

2. Map the Blue Banana zone of development.

3. Trace the boundaries of former Communist Bloc countries in one color and trace the countries of the European Union in another color.

CONSIDER THE DISCIPLINE

CENTRIPETAL AND CENTRIFUGAL FORCES

Slovakian **national sentiment** has survived for hundreds of years, and its rise after the fall of communism hastened the break up of Czechoslovakia.

Former leadership in Eastern Europe denied that ethnic and national tensions were matters of state. Now, **ethnic problems** resurge with vehemence, especially in places such as the Balkan Peninsula in the southern half of Eastern Europe where **racial, religious and regional identities** define people while dividing them. Alexander B. Murphy, a geographer at the University of Oregon, has written extensively on this subject. He notes that most analyses since the fall of the Soviet Union have focused on the prospects for individual European states; yet one of the more important regional divisions within Eastern Europe is much larger in scale, a division between those areas with a long-standing historical association with Western European countries, and those associated with Byzantine and Ottoman Turkish realms. A cultural manifestation of this division can be seen in the **religious geography** of the region; **Western Christianity** dominating the former area, whereas **Eastern Christianity** and **Islam** dominate in the latter. It is significant to note that all three of these religions intersect and vie for influence in the former Yugoslavia.

It is possible to argue that Slovakia, with its history of association with Western Europe, will adjust more quickly than its eastern neighbors to the changing **political geography** of Europe. Yet she will also have to be vigilant during the struggle to develop to guard against making certain ethnic groups targets for blame. The tragedy of Yugoslavia is a lesson in this regard. Efforts not to inflame ethnic and other differences, that is, respecting the ethnic ties that bind nations, will be a key challenge to the new state of Slovakia, which must balance the rapid transformation underway in Europe and the legacy of the former Communist era.

To further your understanding of the field of political geography, consider how supranationalism and devolution can be viewed through the lens suggested by the box on "Centripetal and Centrifugal Forces" in Chapter 1, "Resilient Europe: Confronting New Challenges," from de Blij and Muller, 1994, *Geography: Realms, Regions, and Concepts: 7th Edition.*

CRITICAL VIEWING

Based on the Case Studies in Video Program 3: "Supranationalism and Devolution" from *The Power of Place: World Regional Geography,* and the reading in Chapter 1, "Resilient Europe: Confronting New Challenges," from de Blij and Muller, 1994, *Geography: Realms, Regions, and Concepts: 7th Edition,* develop essays answering the following questions:

1. Discuss the history of the European Union. What are the EU's goals? Is supranationalism critical to Europe's future?

2. Discuss prospects for Czech entry into the European Union.

3. In Western Europe, is supranationalism inevitable? What are the centrifugal forces that threaten realization of a united Europe?

4. In Central and Eastern Europe, is devolution inevitable? What are the centripetal forces that lead to reduced conflict in these regions of Europe?

▪ PROGRAM 4

East Looks West

LESSONS IN GEOGRAPHY

At the end of Program 4, you should be able to:
1. Understand the differing views of the boundaries of the European Realm.
2. Discuss the reasons for and against inclusion of the Russian heartland in the definition of Europe.
3. Understand how political ideology shapes the landscape of the former East and West Berlin.
4. Briefly review the diffusion of democracy in Poland and the reasons for different reactions to it in urban and rural areas.

OVERVIEW

Berlin: United We Stand explores the **land-use characteristics** of a city uniquely shaped by **superpower rivalries** over five decades. Berlin, its division marked by the erection of the Berlin Wall in 1961, is a study in contrasts; differing patterns of land use and **transportation,** as well as of income and residence, can be identified through comparison of the eastern and western parts of the city. The Wall physically enforced a division that began in 1945 in the aftermath of World War II. The Case Study is organized around the efforts of a French geographer to conduct a reconnaissance of selected districts in the former East and West Berlin. Such an inventory would address questions about how and in what ways the dismantling of the Berlin Wall will lead to **integration** of urban areas that formerly were spatially distinct. The resulting changes, in part due to the political designation of the city as the capital of a reunited Germany, and in part a response to international economic forces, represent a dramatic **transformation** of the **urban landscape.** This is especially true along the old boundary separating East and West Berlin, the site of a number of large-scale construction projects.

   Poland: Diffusion of Democracy presents the difficult **transition** from communism to capitalism. Under the new system, Poles are learning the nuances of **discourse, compromise** and **civic engagement.** The future of Poland's democracy depends on how quickly these skills spread thoughout its population. New urban voters have rapidly assimilated these ideas, but rural Poland is plagued by dismal voter turnout. Understanding and incorporating the geographic concepts of **diffusion** — and its **carriers and barriers** — will speed the pace of change in the evolution of this newly **open society.**

   Poland's experiment in democracy and a united Berlin are definitive indicators of how former Soviet satellites will fare in the post-Cold War era.

STUDY RESOURCES

1. Video Program 4: "East Looks West" from *The Power of Place: World Regional Geography*.

2. Unit 2 Map, "Resilient Europe: Confronting New Challenges" from Latz and Gilbert, 1996, *The Power of Place: World Regional Geography Study Guide*.

3. Chapter 1, "Resilient Europe: Confronting New Challenges" from de Blij and Muller, 1994, *Geography: Realms, Regions, and Concepts: 7th Edition*.

PREVIEWING QUESTIONS

1. Where is Berlin?

2. How did the Berlin Wall come to symbolize Cold War rivalries?

3. Where is Poland located in Europe?

4. Where is Berlin located relative to both Western Europe and Eastern Europe?

KEY THEMES

- **A City Divided**
- **Urban Reconstruction – District by District**
- **Reconstruction: European Potential**
- **Poland's Transition to Democracy**
- **Diffusion: A Definition**
- **Grassroots Democracy – One Village at a Time**

DISCUSSION OF CASE STUDY THEMES

• A City Divided

Divided by the allies at the end of World War II, Germany was for decades like a family split into two antagonistic factions, and Berlin itself was marked by this national history. For 28 years, the city had been cut in two by a wall. Then, in 1989, Berlin was reunified and its infamous wall destroyed.

Erected in 1961, for the most part by the Soviet army, the official reason for the wall was to create a barrier to protect the East from its capitalist enemy, the West. It was in fact intended to check escapes from East Berlin to the West. This ominous symbol of the Cold War divided not only a city, but the entire world.

• Urban Reconstruction – District by District

Housing and transportation systems which served neighborhoods with varied characteristics, two distinct centers and one shared but impermeable edge are now challenged by the reconstruction of a single Berlin. Isabelle Aflalo, a French graduate student in geography, has taken an inventory of the spatial variations of the land-use patterns within this newly unified city. With this information she plans to analyze the future direction of metropolitan Berlin.

Aflalo's field work began in the Zoologischer Garten underground station in the Tiergarten District of the former West Berlin. This district's Kurfürstendamm, often compared to the Champs d'Elysées, had faced the East like a showcase of Western riches. Commercial and cultural activities reflect its prior largess. Today, this reunified city must establish a balance where "Ku'damm" will be only one district among many others.

A second district closer to the former dividing line, Kreuzberg, told a different story of land use. While the wall was standing, this area became the refuge of the less fortunate in society, and reflected such diversity with a cosmopolitan lifestyle. A whole culture of cheap, if not free, accommodation characterized the area. A long-standing squatter system has now been legally acknowledged, but Aflalo's research tells her nothing is resolved. Formerly on the edge of West Berlin, Kreuzberg is now one of the most centralized districts and an increasingly desirable area of the united Berlin. Aflalo expects further conflict between the area's traditional squatter and immigrant populations and fashionable newcomers whose presence will propel inevitable rent increases.

Aflalo next visited the Spree River which served as an intermediate zone when the city was split. The zone was no longer in the West, but not quite in the East, since the border was officially designated by the river. Once a no man's land, the land-use plans for this zone are now central to Berlin's future.

Entering the former East Berlin meant leaving subways for tramways along the avenues. These avenues, now in decline, follow a much different plan that those Aflalo left on the west side of the city. The urban layout of the east side is radial rather than linear. In this more classical, controlled urban model, the avenues converge on Alexanderplatz where the historic center of town was once the symbolic center of communist power. An alternative culture now gravitates toward this former center — perhaps a new Kreuzberg.

One refugee fleeing high rent in Kreuzberg told Aflalo that the building he has relocated to in the former East Berlin has 70 disputed claims of ownership. Historically, that building in Friedrichshain belonged to Polish Jews who were thrown out in 1933 by Nazi soldiers. The German Democratic Republic had owned the building for decades, but current residents now claim ownership. In the post-communist reorganization, what constitutes ownership is now a question for the courts.

• Reconstruction: European Potential

Berlin is becoming a reunited city under the influence of the pacesetting and dynamic residents of districts such as Friedrichshain and Kreuzberg. But another important factor in this change comes from the decision to make the city once again Germany's capital. Berlin is preparing to accommodate the headquarters of the Federal Government. As Aflalo was guided through the city, her Spree River tour guide pointed out the new construction and refurbishment at work on future sites of influence — the Chancellory and the parliament building known as the Reichstag.

From the back of a taxi cab, Aflalo's reconnaissance included Potsdamerplatz. This once lively center is poised to reemerge as an economic and financial stronghold since it is situated not only at the center of Berlin, but at the center of a European community potentially enlarged by the addition of many eastern European countries. Expanding on this perhaps not too futuristic hypothesis, Berlin could become an important crossroads. This function, in addition to its position as capital, would allow Berlin to become a thriving political and economic metropolis linking East to West.

• Poland's Transition to Democracy

In 1989, democracy reached Poland and since then the other Eastern Bloc nations of the former Soviet empire. Now these nations face the turmoil of political and economic transition, and their fledgling democracies are being put to a severe test. Transition to democracy has not brought the prosperity that many Poles expected. A recent survey claimed that 70 percent of the population felt life was harder now than it had been under communism.

More than just voting for politicians, Poles are beginning to learn that democracy's decision-making process must permeate governmental, civic, social, and even business organizations.

• Diffusion: A Definition

Diffusion is the spread of an idea, an innovation, a disease — virtually anything — from its source outward across the landscape.

Politically, a country can become "democratic" literally overnight, but true democratic practice spreads slowly and unevenly. About 40 percent of Poland's population lives in small towns and villages, not in cities where carriers such as media, universities, and tourists help spread new ideas and opportunities. Change comes hard in the outlying rural areas where barriers such as isolation and massive unemployment are high, and the carriers — means to transmit new ideas — are few.

• Grassroots Democracy - One Village at a Time

Low voter turnout, the lack of non-governmental organizations, and the absence of an independent press are signs that democracy is lagging badly in Poland's small towns and villages. The Local Democracy in Poland Program is one government effort to introduce democratic practice to this population.

Twenty five towns on the periphery of Poland have been chosen as bases in which to plant the seeds of democracy. Of the 90 trainees schooled in democratic skills at the Warsaw headquarters of the Foundation in Support of Democracy, two were sent as carriers of model democratic behavior back to their remote village of Korsze. In this village of 3,200, record unemployment of 32 percent after the closure of traditional communist cooperative farms has led to suspicion of reformers. Accordingly, the steps to real democracy are small. Instead of starting with a difficult problem, the trainees choose a Ping-Pong tournament as a town's first simple act of citizen-initiated social organization.

If the trainees' model grassroots organizing is successful in Korsze and the other 24 towns, public engagement in community life is expected to spread outward from these hubs to the adjoining regions. Although this diffusion will speed or slow according to local social and economic factors, its organizers believe that democracy is a process that will be hard to stop.

CASE STUDY CONNECTIONS

1. Both the former East Germany and Poland are undergoing a wrenching economic and social transformation from communist to free societies.

2. Both Berlin and Poland act as barometers in Europe regarding the challenges and prospects facing those in favor of democratic rule.

3. The availability of capital to promote the political and economic transformation of former communist states is high in the case of Germany, and low in the case of Poland.

4. Young people will play a critical role in the diffusion and, ultimately, the realization of democratic reforms in eastern Germany and Poland.

5. The contrasts between the social landscape of East and West Berlin, and between urban and rural Poland, reflect the geographic realities that challenge political and business leaders in favor of change and reform in each place.

GEOGRAPHERS AT WORK

One of the chief architects of Poland's plan to spread democratic practice is Polish-born geographer Professor Joanna Regulska of Rutgers University. Regulska emigrated to the United States in 1977. Since 1989, however, she has worked closely with the Polish Parliament to develop the framework for political and economic reform in her former homeland.

Regulska is building on several decades of research which finds people's receptivity to new ideas varies between and within communities. Understanding this spatial variation of diffusion will be crucial to the success of Poland's reforms. As a geographer, Regulska knows that the diffusion of democracy will depend on both the barriers and the carriers.

TEST YOUR UNDERSTANDING

Questions

1. What is the concept of diffusion?

2. How does the study of urban land-use patterns help us understand the challenges facing those who wish to unify Berlin?

3. What are the employment and residential problems that accompany the adoption of free market economies in the former East Germany and Poland?

4. What are the new symbols of unity for Berlin, a city recently divided by an infamous symbol of disunity, the Berlin Wall?

5. What role does Poland's youth play in the diffusion of democratic ideas?

6. Why does the assumption of personal responsibility for one's position in society appear to be so challenging in non-democratic countries?

Map Exercises

The maps provided at the beginning of this unit can be used to complete this exercise. You may find extra copies of these maps helpful. Also, refer to the maps of Chapter 1, "Resilient Europe: Confronting New Challenges," from de Blij and Muller, 1994, *Geography: Realms, Regions, and Concepts: 7th Edition*, and any standard atlas to assist in this exercise.

1. Place the following names:

Poland	Warsaw	Baltic Sea
Korsze	Krakow	Gdansk
Tarra Mountains	Belarus	Ukraine
Lithuania	Kaliningrad	Czech Republic
Prague	Bohemian Basin	Germany
Berlin	Spree River	Carpathian Mountains
Elbe River	Sudeten Mountains	Austria
Denmark	Paris	

CONSIDER THE DISCIPLINE

EUROPE: THE EASTERN BOUNDARY

One would expect geographers to be in agreement about what constitutes the world's great land masses and its oceans and seas. When it comes to defining Europe as a physical and cultural entity, however, differences of opinion need to be balanced and explained.

It is generally agreed that the **European realm** is bounded by bodies of water on the west, north, and south, respectively the Atlantic, Arctic, and Mediterranean. Land areas that exist on the periphery of Europe include the island countries of the United Kingdom and Iceland to the northwest, Scandinavia to the north, the Andalucian region of Spain to the southwest, and the Greek Islands to the southeast. Defining Europe's eastern boundary, however, is a matter of some dispute.

Three options exist for classifying **Europe's eastern boundary:** the first is the eastern borders of Poland, Slovakia, Hungary and Romania; the second is the eastern borders of Latvia, Lithuania, Belarus, and Ukraine; and the third is the Ural mountains of Russia, a territory that stretches 992 miles (1,600 kilometers) beyond Moscow. The companion text for this course, de Blij and Muller, 1994, *Geography: Realms, Regions, and Concepts: 7th Edition,* subscribes to a definition of Europe that coincides on the east with Russia's western border. The European advisors to *The Power of Place,* on the other hand, define Europe as including a region of Russia extending to the Ural mountains.

The argument hinges on whether there is a region within Russia that can be characterized as having European qualities due to a long history of interaction, that is, the **Russian heartland** west of the Urals, or whether the great country of Russia deserves to be in a class by itself based on its political geography and large territorial size and population.

It would be sheer folly to see these differences of opinion as merely "academic." The differences of opinion about what constitutes Europe influences the long-term potential of a **European Union** — a kind of United States of Europe — and whether it is to be defined in terms of political and economic values, or whether physical territorial unity and pre-Soviet Union historical relations among Europe's countries will also be considered. And much of the contemporary research on the **environment** documents with chilling detail that political divisions between states in Europe cannot be allowed to interfere with the pollution problems that seep across these human territorial divisions that **nation-states** represent.

To further develop your understanding of this concept, refer to the box "Europe: The Eastern Boundary" in Chapter 1, "Resilient Europe: Confronting New Challenges," from de Blij and Muller, 1994, *Geography: Realms, Regions, and Concepts: 7th Edition.*

CRITICAL VIEWING

Based on the Case Studies in Video Program 4 "East Looks West" from *The Power of Place: World Regional Geography,* and the reading in Chapter 1, "Resilient Europe: Confronting New Challenges," from de Blij and Muller, 1994, *Geography: Realms, Regions, and Concepts: 7th Edition,* develop essays answering the following questions:

1. Discuss the history of the Berlin Wall, the circumstances that brought about its establishment, and the regional factors, particularly in the former Soviet Union, that led to its dismantling.

2. Uncertainty about the future is a common denominator facing the citizens of formerly Communist East Berlin and Poland. Do you think worries about the future among former East Berliners and Poles is more or less pronounced than that in the U.S.?

3. The possible reemergence of Germany at the center of a united Europe is cause for celebration for some, grave concern for others. Account for these differing views. In the case of those who worry about the central position that Germany may assume, what steps might its government take to address such concerns?

4. Suppose that Poland reverted to communism and utterly rejected its experiment in free markets and a democratically elected government. How does geography help to explain this proposition? And how might the prospect of a reemerging communist state be greeted by Poland's European neighbors?

5. What policy efforts might the United States make to encourage the adoption of free market principles in East Germany and Poland? Are the steps you propose economic, political, humanitarian, or educational? Be specific in the formulation of your recommended plan of action.

■ PROGRAM 5

The Transforming Industrial Coreland

LESSONS IN GEOGRAPHY

At the end of Program 5, you should be able to:

1. Discuss how the location of Liverpool aided in its development as a center for trade.
2. Explain the role of government in industry and trade development in the Netherlands and England.
3. Describe how the development of high-speed rail may affect the lives of the residents of Liverpool and the Randstad.
4. Explain the basis of the opposition to high-speed rail in the Randstad.

OVERVIEW

Program 5 highlights two places in Europe in *Liverpool: A New Dawn* and *Randstad: Preserving the Green Heart*. This city in England and this metropolitan region of the Netherlands are tied together by the common themes of **modernization, transportation, trade**, and concern for the **quality of life.**

Liverpool was settled in 1207 in northwestern England, after King John granted a charter for a new, planned town on the northeastern shore of the Mersey Estuary. The city developed during the mid-17th century as the main **port** linking England with Ireland. In the 17th and 18th centuries, the port handled colonial trade and during the Industrial Revolution served the nearby **manufacturing** complex centered in Manchester. Now an industrial city itself, Liverpool's city center is 3 miles (5 kilometers) from the Irish Sea, but docks extend for 5 miles (8 kilometers) northward along this flat coast. By the 1970s, the port had begun to lose business because of its lack of modern equipment which would speed the loading or unloading of goods. When Liverpool's docks were modernized and **containerized cargo** became predominant, many stevedores who had worked the docks became unemployed. With the completion of the **Channel Tunnel (Chunnel)** link to the European continent and improvements in technology, Liverpool is making a comeback.

The **Randstad** megalopolis is anchored by three cities — Amsterdam, Rotterdam, and The Hague — which itself forms the cornerstone of the Netherlands' core area. As explained in the de Blij companion text to this telecourse, this **conurbation** and its coalescence has created "a ring-shaped, multi-metropolitan complex that surrounds a still-rural center (the literal translation of *rand* is edge or margin). A more precise labeling of the conurbation, however, would be Randstad-Holland, because *Holland* (meaning "hollow country") refers specifically to the Dutch heartland that faces the North Sea in these lowest-lying western provinces of the Netherlands." *Randstad: Preserving the Green Heart* explores the protests of residents and farmers in the still-agricultural pockets of this region who wish to preserve this region's remaining "green" character in the face of **development**. Currently, the government backs a plan to develop **high-speed rail** access to Schiphol airport from Rotterdam, and the most efficient route cuts right through the green heart of the Randstad. The people of the Randstad see the need for a link with the rest of Europe via Rotterdam, but those interviewed feel this area needs to preserved. The government sees no way to both bypass the green heart and maintain Schiphol's **comparative advantage** as a **transportation center** providing access to greater Europe via both air and land.

STUDY RESOURCES

1. Video Program 5: "The Transforming Industrial Coreland" from *The Power of Place: World Regional Geography*.

2. Unit 2 Map, "Resilient Europe: Confronting New Challenges" from Latz and Gilbert, 1996, *The Power of Place: World Regional Geography Study Guide*.

3. Chapter 1, "Resilient Europe: Confronting New Challenges" from de Blij and Muller, 1994, *Geography: Realms, Regions, and Concepts: 7th Edition*.

PREVIEWING QUESTIONS

1. Where is Liverpool located in the United Kingdom?

2. What is the scope of the area considered to be the hinterland of Liverpool?

3. Where is the Randstad connurbation? What is its relative location to Germany and France?

4. Where does the Chunnel link England to continental Europe?

KEY THEMES

- **Modernization**
- **Transportation**
- **Trade**
- **Quality of Life**

DISCUSSION OF CASE STUDY THEMES

• Modernization

Both Liverpool and the Randstad are concerned with the effects of modern development. Each community's experience with modernization, however, has been quite different. Liverpool is suffering from outdated industrial equipment and is struggling to modernize. Its port facilities have been in place for many decades but are far behind the world's current technological standards. This lack of modern equipment led to decreased maritme trade and prompted shipping operations to relocate, resulting in a high rate of unemployment. Installation of new equipment has begun to attract the return of business, but has not, however, restored the employment opportunities known in Liverpool's heyday. Today, in addition to updating the port facilities, Liverpool is also exploring land-based alternatives to foster its continued growth.

The Randstad's frustrations contrast those of Liverpool. Some in this area seek to prevent the effects of modernization from further altering the lives of its residents. The government wishes to pursue policies aimed at constructing and operating a high-speed rail line between Schiphol Airport and Rotterdam, but residents of the Randstad protest such development in their green space.

• Transportation

The means by which goods are moved varies from air to sea to land. Transportation, especially land-based modes, drives the economies of many cities. The ways in which cities cope with the modernization of transportation often predict whether they will thrive in the future. At present, Liverpool struggles to catch up with other cities, while the Randstad seeks to avoid the consequences of modernization. One common thread between these places is the advent of high-speed rail. Liverpool has the opportunity to take advantage of this new technology, while some resident of the Randstad seek to avoid being affected by it.

Liverpool's boom came in the 1950s when the port served as one of the most important gateways in Europe. In the 1990s, Schiphol Airport strives to expand its role as a leading European gateway. The decisions each community makes will be pivotal to its future.

• Trade

Trade among nations is the driving force determining the futures which Liverpool and the Randstad face. The process of trade fuels the development of transportation networks among areas to speed the flow of goods, people, and ideas. The need to move products and people with increased speed and efficiency leads to the modernization of facilities to meet demands. Failure to modernize may cause the loss of jobs in an area which is not able to, or not willing to, plan for the future. The links which are created to encourage trade may impact the lives of people living along transportation routes.

• Quality of Life

The people in Liverpool have felt the pains of extended unemployment and hope concerted efforts of government and business leaders will bring prosperity. The people of the Randstad have enjoyed a quality of life for many years that is now threatened by the forces that attempt to further modernize its transportation sector. Government policies in both areas will strive to mediate these competing concerns.

CASE STUDY CONNECTIONS

1. Liverpool and the Randstad are gateways to the rest of Europe's core.
2. The concept of relative location contributes to both the Randstad and Liverpool in their cycles of growth and decline as centers of trade.
3. Liverpool developed as a trade hub; the Randstad developed first through the production of food, then as an important international trade center.

4. The integration and reprioritization of sea-, air- and land-based transportation modes are at the heart of the development plans of both areas.

GEOGRAPHERS AT WORK

One of the areas of concentration within the field of geography is urban geography, the study of the city: its genesis, location, form, and evolution over time. Peter Lloyd, a Professor of Geography at Liverpool University, is one such urban geographer.

Lloyd's interests concentrate on the cycle of change affecting Liverpool. Some of the forces giving rise to Liverpool's current circumstances are global; for example, its rise and decline as a major world port parallels that of the British Empire. Another factor that informs his study of the city is technology, and the capacity of city leaders and dock workers to adapt to changing port functions as containerization of cargo reduced the demand for labor servicing maritime trade companies. The consequences of global and local change affect the physical and functional structure of Liverpool as a city and as a port. This is the background to Liverpool's contemporary urban geography.

TEST YOUR UNDERSTANDING

Questions

1. The location of each Case Study has influenced its functional specialization. Places like Liverpool, located on the coast, have the opportunity to develop as ports for trade or fishing. What factors in the past have influenced Liverpool to develop as a port for trade?

2. The process of modernization can have effects which people either appreciate or dislike. Describe these effects and people's reactions in both Liverpool and the Randstad.

3. Discuss several of the reasons trade occurs between nations. *speed flow of goods, people, ideas*

Map Exercises

The maps provided at the beginning of this unit can be used to complete this exercise. You may find extra copies of these maps helpful. Also, refer to the maps of Chapter 1, "Resilient Europe: Confronting New Challenges," from de Blij and Muller, 1994, *Geography: Realms, Regions, and Concepts: 7th Edition*, and any standard atlas to assist in this exercise.

1. Place the following names:

United Kingdom	England	Scotland
Pennine Mountains	Mersey Estuary	Wales
Irish Sea	Dublin	Irish Republic
Northern Ireland	St. George's Channel	London
English Channel	Strait of Dover	Manchester
The Netherlands	Amsterdam	Rotterdam
The Hague	Randstad	Belgium
European Core Boundary	North Sea	Germany

CONSIDER THE DISCIPLINE

GEOMORPHOLOGY

The physical geography of Europe is both distinctive and regionally varied. The spatial unity that distinguishes the **North European Lowland** (or Great European Plain) is a case in point. Agricultural diversity, intensity of land use, a plethora of navigable rivers, and high concentrations of population are hallmarks of this European region.

The North European Lowland extends in a vast arc from southwestern France, north and east along the North Sea through the Low Countries of the Netherlands, Belgium and Luxembourg, across northern Germany and then eastward through Poland and into southern Russia. The Lowland designation also includes southeastern England, Denmark, and the southern tip of Sweden.

Of all the European countries found in the North European Lowland, perhaps none typifies it better than the Netherlands. Within the Netherlands, the Randstad, the triangular core conurbation consisting of Amsterdam, Rotterdam, and The Hague, illustrates the geomorphic background for one of the world's unique contemporary landscapes.

Two great rivers reach the sea in the Netherlands, the **Meuse (Maas)** and the **Rhine**. The vast and ancient **delta land of swamps,** created where each river enters the **North Sea,** have been reclaimed since the early Renaissance through extensive and ingenious application of **hydraulic engineering technology** consisting of dams, dikes, and polders. Much of the country would be a shallow sea bottom were it not for these efforts, and it is not too great a generalization to observe that the entire landscape of the Netherlands is a result of human effort. By virtue of these efforts, extremely productive **farmland** has resulted for the growing of flowers, hay and grasses, as well as dairy foods. Europe's busiest port, Rotterdam, is now the **shipping gateway** for Western Europe, and Amsterdam's international airport, Schiphol, is one of the most important on the continent as measured by movement of passengers and air cargo.

Much like Singapore in Southeast Asia, and Japan in East Asia, the Netherlands has overcome the **limitations of its physical geography**, namely space and paucity of resources, to become preeminent in international trade, and the country stands poised to continue that tradition as the 21st century approaches.

To further develop your understanding of this field of regional geography, refer to the Focus on a Systematic Field in Chapter 1, "Resilient Europe: Confronting New Challenges," from de Blij and Muller, 1994, *Geography: Realms, Regions, and Concepts: 7th Edition*.

port in Rotterdam

airport in Amsterdam

CRITICAL VIEWING

Based on the Case Studies in Video Program 5 "The Transforming Industrial Coreland" from *The Power of Place: World Regional Geography,* and the reading in Chapter 1, "Resilient Europe: Confronting New Challenges," from de Blij and Muller, 1994, *Geography: Realms, Regions, and Concepts: 7th Edition,* develop essays answering the following questions:

1. Has your town ever felt the influence of unemployment like Liverpool has? Discuss what possible factors may influence unemployment in your community. If your town has been experiencing growth, what has caused it?

2. The need to modernize transportation occurs on a continuing basis. Discuss how your community benefits from its transportation system. Is there widespread consensus for continued investment in the development of such infrastructure?

▪ PROGRAM 6

Challenges on the Periphery

LESSONS IN GEOGRAPHY

At the end of Program 6, you should be able to:

1. Identify the location of Iceland and discuss its cultural and physical relationship to the rest of Europe.

2. Describe the location of Andalucía within Spain and the reasons for its current economic recession.

3. Discuss the conflicts now found in Iceland regarding fishery resource management.

4. Understand the prospects for Andalucía in its struggle to integrate itself with Europe's regional core.

OVERVIEW

Iceland: The Edge of the Habitable World examines the many challenges this island micro-state faces due to its **remote location** and **harsh environment,** as well as its cultural evolution on the **periphery** of Europe. The Case Study focuses in particular on the fishing industry, and the reasons behind its critical, and currently threatened, role in Iceland's economy.

Andalucía: Life on the Periphery explores the once-prosperous Andalucía region of Spain, and the gradual **marginalization** of this area, now peripheral to Europe's **industrialized core**. A geographer studies the ways in which Expo '92 spurred hopes that another "Silicon Valley" would take root in the Expo's now-vacant complex near Seville and revitalize this region gripped by **economic recession.**

STUDY RESOURCES

1. Video Program 6: "Challenges on the Periphery" from *The Power of Place: World Regional Geography*.

2. Unit 2 Map, "Resilient Europe: Confronting New Challenges" from Latz and Gilbert, 1996, *The Power of Place: World Regional Geography Study Guide*.

3. Chapter 1, "Resilient Europe: Confronting New Challenges" from de Blij and Muller, 1994, *Geography: Realms, Regions, and Concepts: 7th Edition*.

PREVIEWING QUESTIONS

1. Where is Iceland?
2. What historical factors lead Iceland to be classified as a European country?
3. Indentify the location of Andalucía within Spain.
4. What was Expo '92? *World fair celebrating 500th anniv discovery of America*

KEY THEMES

- **Human/Environment Interaction in Iceland**
- **Resource Management on the Periphery**
- **Andalucía: Once the Core, Now the Periphery**
- **Economic Geography**

DISCUSSION OF CASE STUDY THEMES

• Human/Environment Interaction in Iceland

Icelanders depend on a hostile environment for their survival. The ocean fisheries surrounding Iceland supply 80 percent of the country's exports, bringing wealth to many and fueling Iceland's strong service economy. However, the northern Atlantic Ocean which brings fortune can also bring misery when frequent wind, rain and blizzards create hazardous conditions. Since 1860, in one small town on Haemay Island, 500 fishermen of 5,000 inhabitants have lost their lives in the stormy seas.

Iceland's stormy seas, however, have provided the abundant seafood stocks which support this island's high standard of living. Here problems and prospects are a function of physical geography. Powerful ocean dynamics — the meeting and mixing the warm Gulf Stream and an ice-bearing Arctic current — are at play in the shallow waters of the Icelandic coasts. This mixing of warm and cold creates an upwelling of nutrient-rich water feeding a strong fishery.

Iceland is literally located on the edge of the North American and Eurasian tectonic plates. The continuous separation of these plates along the Mid-Atlantic Ridge makes Iceland one of the most volcanically active places in the world. Millions of years ago, volcanoes on the ocean floor erupted to create the islands now called Iceland. Lava and waves created one of the country's finest natural harbors on Haemay, providing shelter in the midst of Iceland's most productive and most dangerous waters.

• Resource Management on the Periphery

Despite the bounty brought about by Iceland's physical geography, Icelanders now face a fish shortage. The technologically advanced fleets of these islands have depleted the fishery resource.

Scientists believe that with careful regulation of this technology, the fish stocks could be stabilized. The government has imposed and monitors quotas on the size of every fisherman's catch. But the struggle to find balance between the health of resource and financial concerns of the fishermen continues to be debated.

technologically advanced fishing fleet has cleaned out the fish.

Whether meeting the challenges of the earth or the sea, Icelanders have always lived on the edge. Their future depends on their ability to overcome the disadvantages of their physical geography and maintain a foothold in this unique physical environment.

• Andalucía: Once the Core, Now the Periphery

Situated at one meeting point of the Islamic, Jewish and Christian worlds, Andalucía was central and powerful during the Middle Ages. And it became, in the 17th century, the turntable of exchanges between the Mediterranean and Atlantic worlds. Thus, Seville, the region's capital, has conserved in its monumental architecture the flamboyance of a period which saw all the riches of a new world pass through its gates. Subsequently, the Europe of conquistadors gave way to that of the industrialists. Despite its mineral resources, the region suffered from the inertia of the ruling class and was left behind by the Industrial Revolution. Spain became progressively more marginal in Europe, and Andalucía more marginal in Spain.

• Economic Geography

Less than 15 years ago, almost all of the Andalucian coastline was just sand, palm trees and tourist homes. Today, thousand of new inhabitants coming from Spain's interior have colonized these plains under a sea of plastic greenhouses. Many new highways transport tomatoes and other fresh food products directly to the great consuming markets of the European core. In this way, Andalucía could be regarded as an integrated periphery, but leaders in the region want to be more than an orchard for Europe.

While maintaining its prodigious tourist economy and certain agricultural advantages, Andalucía wants to attract different sectors of commerce — industry, technology, and services. Initially conceived as a world fair commemorating the 500th anniversary of the discovery of America, the Expo "Cartuja '92" was, in the end, dedicated to the celebration of the new frontiers of science and technology. After the Expo ended, it was hoped that its new buildings would attract high-tech industries. Even with new transportation links to the European core, planners found that it is not so easy to reproduce on command the experience of California's Silicon Valley, or of other places where technological development goes hand in hand with a good quality of life.

CASE STUDY CONNECTIONS

1. Both Andalucía and Iceland are located on the periphery of Europe.
2. Both depend on the careful management of resources as mainstays of the economy.
3. One of the environmental challenges facing Iceland, severe cold, can be contrasted to the Mediterranean environment of Andalucía.
4. The history of each country differs greatly; Andalucía was once central in Europe while Iceland has always been peripheral to it.
5. The role of government in each country will be critical to providing the vision, infrastructure investment programs, and economic policies to maintain much less improve current and future living standards.

GEOGRAPHERS AT WORK

Andalucía: Life on the Periphery features Kattalin Gabriel, a geographer studying the distinctive regional geography of Spain, a country famous for its cultural, economic, and physical differences from north to south, east to west.

It is common in regional studies to ask questions about how places develop. But here an equally important question is raised; why is it that some areas of a country appear to be bypassed in the course of economic and political development? A case in point is Andalucía, one of the poorest regions in Western Europe. This matter is all the more striking given the historical importance of the Andalucian region.

As Gabriel discusses in the video, much of the explanation for the Andalucía's fall from grandeur has to do with its relative location to traditional European and Mediterranean centers of power. Since the Industrial Revolution, Europe's core has shifted inexorably northward, so that Andalucía now finds itself on the periphery. Policies to integrate this area into Europe's orbit are a source of hope, illustrated by the strategic investments occurring in high-speed transportation. Nonetheless, the capacity of Andalucía to create a diversified and robust economy in the future remains very much in doubt.

TEST YOUR UNDERSTANDING

Questions

1. Why did the Andalucian government place such high hopes that sponsorship of Expo '92 would lead to a revitalization of the regional economy?

2. Explain how technology has altered the human/environment relationship in Iceland. In Andalucía.

3. What steps have the Icelandic and Spanish governments taken to resolve economic difficulties for the people dwelling in each area?

4. Identify the physical and cultural contrasts between Iceland and Andalucía.

Map Exercises

The maps provided at the beginning of this unit can be used to complete this exercise. You may find extra copies of these maps helpful. Also, refer to the maps of Chapter 1, "Resilient Europe: Confronting New Challenges," from de Blij and Muller, 1994, *Geography: Realms, Regions, and Concepts: 7th Edition*, and any standard atlas to assist in this exercise.

1. Place the following names:

Atlantic Ocean	Arctic Circle	Iceland
Reykjavik	Haemay Island	Norwegian Sea
European Core Boundary	Mediterranean Sea	Tropic of Cancer
Spain	Andalucía	Seville
Madrid	Strait of Gibraltar	Algiers

CONSIDER THE DISCIPLINE

GEOMORPHOLOGY

As seen in *Iceland: The Edge of the Habitable World,* the earth's crust is constantly under assault by natural forces. Its landform surfaces are continually being shaped by the processes of weathering and erosion, and the very structure of the earth, including vast areas not visible below the surface of the oceans, is shaped by internal crustal movements known as **tectonic forces.** Within the field of geography, those who study the configuration of the surface of the earth are called **geomorphologists** and their field of study, geomorphology.

The field of geomorphology sets out to explain the spatial diversity of the earth's physical landscapes. Its students have contributed brilliant insights into how the world is structured; among the most famous examples is the work nearly a century ago of **Alfred Wegener** who inspired a number of geographers and geologists with his ideas about how the continents came to be located on the earth's surface in their current configuration. Wegener's theory is the background to the process of **continental drift,** the movement over the last several hundred million years of whole landmasses relative to one another across the surface of the planet.

It is now widely accepted that the earth's **crust** consists of tectonic plates floating atop a thick, molten layer called the **mantle.** But it is only over the last four decades that the evidence to support this theory, originally presented through a comparison of physiography, geology, and paleontology of continental features in such places as Africa and South America, has come to be part of the canon of scientific thinking. Bold thinking by Wegener and his students postulated that the continents were once linked in a supercontinent called **Pangaea** whose breakup could be traced through the theory of continental drift. The critical evidence confirming this theory, interestingly, came not from the continents themselves, but from the ocean floor.

As knowledge of the seafloor's characteristics has increased over the course of the 20th century, scientists have concluded that its dominant physical feature is a globe-circling **submarine mountain chain** known as the **Mid-Oceanic Ridge.** Research has confirmed that this ridge is the by-product of massive amounts of upwelling **magma** from the mantle. This outpouring of molten material, or lava, creates mountains and canyons; at the ridge's center, the youngest geological rocks can be found; at its extremities, the oldest. The seafloor is literally spreading due to this ridge formation, most rapidly in the southeastern Pacific, where the ridge is broad and low, and most slowly in the mid Atlantic, where it is more mountainous.

Iceland sits atop the **Mid-Atlantic Ridge,** its **topography** mirroring the submarine formations of this mountain system. Thus, its scale may be explained in terms of forces that are truly global in scope. At the same time, when considered in terms of its location on the northwestern periphery of Europe and just south of the Arctic Circle, this island country of active volcanoes and severe winters could not support a population of more than one-quarter million people without carefully managing its natural resources. Iceland's economic activity is almost entirely oriented toward **offshore waters** traditionally **rich in sealife** but now undergoing rapid transformation due to **overfishing.**

To further develop your understanding of this field of regional geography, refer to the Focus on a Systematic Field in Chapter 1, "Resilient Europe: Confronting New Challenges," from de Blij and Muller, 1994, *Geography: Realms, Regions, and Concepts: 7th Edition.*

CRITICAL VIEWING

Based on the Case Studies in Video Program 6: "Challenges on the Periphery" from *The Power of Place: World Regional Geography,* and the reading in Chapter 1, "Resilient Europe: Confronting New Challenges," from de Blij and Muller, 1994, *Geography: Realms, Regions, and Concepts: 7th Edition,* develop essays answering the following questions:

1. The circulation of people, goods, and ideas is one of the keys to economic development. Business and government leaders support the building of a high-speed rail link between Madrid and Seville. Would you advise the Andalucian government to continue similar transportation investments in the 21st century?

2. Iceland possesses significant geothermal energy resources. Devise a research agenda for assessing the value of these sources of energy both for Iceland and for other countries of the world.

3. Icelanders, at least based on the video, appear not to understand the seriousness of the problems created by overfishing in their local waters. Devise an educational curriculum, aimed at both children and parents, that addresses this lack of awareness.

4. International trade between Andalucía and European countries to the north of Spain has potential but is full of challenges. Is it possible that other international trade connections, with North Africa, the Middle East, or North America, might represent options? Explore this question.

▪ UNIT 3
Russia's Fracturing Federation

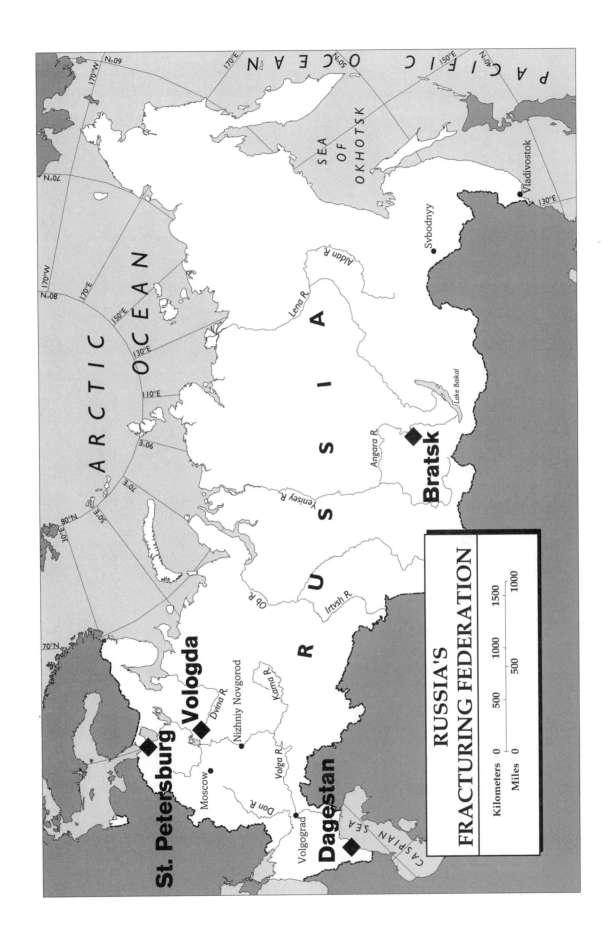

RUSSIA'S
FRACTURING FEDERATION

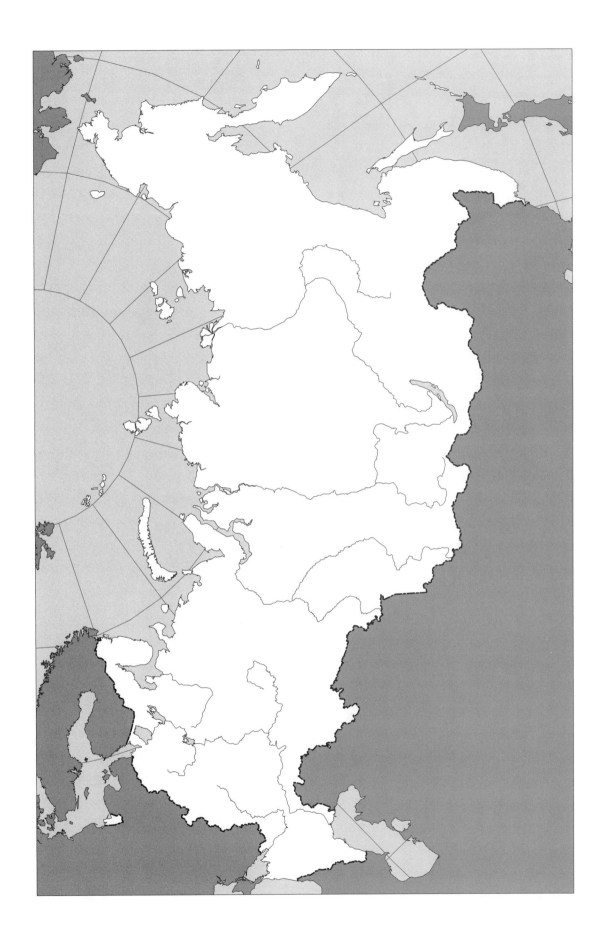

▪ PROGRAM 7

Facing Ethnic and Environmental Diversity

LESSONS IN GEOGRAPHY

At the end of Program 7, you should be able to:

1. Describe how the differing cultures in Dagestan affect the republic.
2. Identify the cultural factors which have promoted political conflict in the Caucasian Periphery.
3. Explain why great cultural differences can exist among regions within Russia.
4. Illustrate how the expansion of Islam has caused political and cultural changes in southwestern Russia.
5. Identify the distribution of different types of climate, soils, and resources in the Russian Core Area and the Caucasian Periphery.
6. Characterize the constraints of the climate and soil in Vologda.
7. Evaluate the environmental limitations of natural resource-based industry in Vologda.

OVERVIEW

Since the **dissolution** of the vast Soviet Union, **nationalities** of the republics have been confronted with a choice between nationalist anarchy or peaceful cooperation with the Russian Core. Among the dry reaches of the Caucasus Mountains lies the Republic of Dagestan, comprised of over 30 ethnic groups. In *Dagestan: Russia's Southern Challenge,* we see how a complex tapestry of religion, culture, history and economy unifies this Republic's people who are Russian by rule but not background.

In *Vologda: Russian Farming in Flux,* we see the quintessential Russia. This region, surrounding the administrative center bearing its name, is **remote** and **rural**. Many of the changes which have occurred since the collapse of the Soviet Union have not yet reached Vologda. There, **collective farms** operate in much the same fashion as they did under communist rule.

The contrast between the peoples of Dagestan and Vologda illustrates the variety of thought and behavior found in Russians from places far apart yet under the same federal rule. The diversity of their problems stems in large part from their disparate locations, climates, and histories — their geography — within this vast country. Just as their past has been shaped by these factors, the solutions that will occur to them and seem suitable are strongly influenced by background, culture, traditional beliefs, and environmental limitations.

STUDY RESOURCES

1. Video Program 7: "Facing Ethnic and Environmental Diversity" from *The Power of Place: World Regional Geography*.

2. Unit 3 Map, "Russia's Fracturing Federation" from Latz and Gilbert, 1996, *The Power of Place: World Regional Geography Study Guide*.

3. Chapter 2, "Russia's Fracturing Federation" from de Blij and Muller, 1994, *Geography: Realms, Regions, and Concepts: 7th Edition*.

PREVIEWING QUESTIONS

1. Where is Dagestan? Which body of water is to the east? Which republic is to the southwest?

2. Where is the city of Vologda within the region of Vologda?

3. What is the hierarchy of the Russian republics, autonomous regions, and the federation?

3. Explain the role of collective farming in the centrally planned economy of the former Soviet Union.

KEY THEMES

- **The Distinctive Geography of Dagestan**
- **Uniting Factors in the Caucasian Periphery**
- **Relationship with Russia**
- **Physical Constraints in Vologda**
- **Resource-Based Economy**
- **Collective Transition**

DISCUSSION OF CASE STUDY THEMES

• The Distinctive Geography of Dagestan

In mountainous Dagestan, every village, or *aoel*, is a small society unto itself, with its own dialect and cultural traditions. More than 30 ethnic groups or nationalities — each with its own traditional dress, music and dance — live in this Russian republic in the northern Caucasus range. The physical features of these mountains serve to reinforce ethnic distinctions. The small mountain villages are spatially, and thus culturally, isolated from one another.

• Uniting Factors in the Caucasian Periphery

With its location between Russia and Southwest Asia, Dagestan has been settled by a variety of peoples. Here at the northern edge of the spread of Islam, Russia and religion give these otherwise disparate groups a common link. As stories told in the video reveal, resisting Russian colonization but then eventually sharing Russian as a second language are but two ways the different peoples of Dagestan are the same.

new roads
+ tunnels
paid for
by Moscow

• Relationship with Russia

Russians fought the Persians for Dagestan as early as the 15th century. Not until four centuries later when the Caucasian War ended in 1877 did local resistance fall to Russian control. With little in-migration Russian ethnicity has reached only 11 percent since Dagestan was named an Autonomous Republic in 1921.

One example of Russia's economic interest in Dagestan is dam construction. When Russia uses the republic for hydropower development it leaves behind the effects of new infrastructure, electricity, and employment opportunites which are helping Dagestan to modernize.

Now, Dagestan remains united, watching the troubles of its separatist neighbor, Chechnya, as it gains an economic advantage by retaining ties with Russia.

• Physical Constraints in Vologda

poor soils, sandy

The northern latitude of the region and city, both known as Vologda, and their inland locations, explain long, cold winters. The region's continentality prohibits the moderating effects of ocean currents on extreme temperatures, and Vologda's winter low temperature can drop to -40°F (-40°C).

Vologda's location is far north of the "black earth zone" where Russia's best soils are found. Due to the region's northerly location and sandy and poor soils, the growing season and the productivity of the land are severely limited. Under these difficult physical conditions, dairy farming is a logical specialization.

cold winters
need irrigation

• Resource-Based Economy

Vologda's northern position borders the boreal forest zone. Seventy percent of its surface is forested, and it has a timber industry that exports its products to Sweden, Norway, Finland, and recently, France. Vologda's resource stock and opportunites for processing complement the short crop growing season and facilitate the adaptation to dairy farming.

Value is added to forestry resources through local processing, providing an essential source of income for the region when products such as industrial paper and cardboard are shipped to foreign buyers. Product quality and international trade are new issues concerning the management of Vologda's forestry resource. As seen in the paper processing plant, the philosophy and production methods of the market system are becoming ever more a part of this region of Russia.

• Transition from Collective to Market System

When sales were subsidized under the planned economy, Russian farmers didn't have to worry about year-round productivity or diversifying the use of their resource base. During the Soviet era, farms and other collective enterprises took care of housing for their employees. Farms arranged services such as shops and schools: the farm was actually more like a family village.

Employees still live in the same houses, but, with privatization, some of those houses have now become private property. Under the new market system, people must find some activity beyond farming to support themselves. Competition, efficiency and quality are not only new words but new ways of thinking during the post-Communist transition.

shared
religions

prefer to
be independent
but know they
need Russia

CASE STUDY CONNECTIONS

1. Dagestan is an autonomous republic, and Vologda is a region in Russia, a country that spans 11 time zones.

2. The Caucasian Periphery of the Russian southwest is home to Dagestan; Vologda is part of the Russian Core.

3. The homogeneity of the population of Vologda can be contrasted to the 33 distinct nationalities which comprise Dagestan.

4. Dagestan has sizeable oil and gas deposits, while 70 percent of Vologda is forest.

5. Vologda is far from the sweep of recent political upheaval; Dagestan lies in a sensitive area.

GEOGRAPHERS AT WORK

In *Facing Ethnic and Environmental Diversity*, geographer Andrei Treivish of the Academy of Sciences provides a number of insights into the ethnic geography of Dagestan.

Russia's ethnic groups are diverse, widely distributed, and have created distinctive landscapes of land-use and culture groupings throughout the country. Their interaction is a key to national stability, as illustrated by the conflict in Chechnya, immediately adjacent to Dagestan.

The study of ethnicity in Russia is a major field of endeavor by academic geographers who see such matters as spatial and territorial. Whereas many researchers represent ethnic groups simply as a percentage of the larger population, geographers are especially interested in ethnic groups' distributional patterns. The distribution of people tied by ethnic identity to particular territories holds immense significance for the future of the Russian state, as well as for other countries in the world, especially when these people seek to create autonomous areas of control.

TEST YOUR UNDERSTANDING

Questions

1. What are some unifying influences for the people of Dagestan?
2. What factors influence the size, type, and location of settlements in Dagestan?
3. List the physical constraints faced by the people of Vologda.

Map Exercises

The maps provided at the beginning of this unit can be used to complete this exercise. You may find extra copies of these maps helpful. Also, refer to the maps of Chapter 2, "Russia's Fracturing Federation," from de Blij and Muller, 1994, *Geography: Realms, Regions, and Concepts: 7th Edition*, and any standard atlas to assist in this exercise.

1. Place the following names:

Caucasus Mountains	Caspian Sea	Dagestan
Chechnya	Azerbaijan	Volga River
St. Petersburg	Vologda	Moscow

2. Map the distribution of climate, soils and vegetation across the regions studied.

3. Indicate on a map where the following occur:

> Russian Boundary of 1992
>
> Russian Core Area
>
> Extent of former USSR

CONSIDER THE DISCIPLINE

CLIMATOLOGY

Climatology is one of the main branches of **physical geography.** The term climate implies an average or long-term record of weather conditions at a certain place or region. Climatologists study the **distribution of climatic conditions** over the earth's surface.

The distribution of climatic conditions has a profound affect on the habitability and agricultural capacity of a place, region, or country. Key factors contributing to climate distribution include **latitude** and **continentality.** In general, the farther away a place is from the equator, the colder its climate. And the farther inland a place is from the moisture and moderating influence of the earth's oceans, the climate will be drier and will have more extreme fluctuations in temperature.

Russia is a high latitude country. Because its topography is relatively flat to the north, Arctic air has free reign from the Kola to the Kamchatka Peninsula. And warm, most air that might have come from the south is blocked by the mountains which surround Russia and its former Soviet Republic neighbors, like Dagestan. As a result, the entire southern region from the Caspian Sea to the mountains bordering China, is a **desert of steppe,** or "a semi-arid grassland."

As a result of Russia's climatic pattern, the majority of its **population** remains **concentrated** in the western fifth of this gigantic country. Russia's **high latitude continental climate** is the reason why agriculture remains a critical problem, as seen in the Vologda Case Study.

To further develop your understanding of this field of physical geography, refer to the Focus on a Systematic Field in Chapter 2, "Russia Fracturing Federation," from de Blij and Muller, 1994, *Geography: Realms, Regions, and Concepts: 7th Edition.*

CRITICAL VIEWING

Based on the Case Studies in Video Program 7 "Facing Ethnic and Environmental Diversity" from *The Power of Place: World Regional Geography,* and the reading in Chapter 2, "Russia's Fracturing Federation," from de Blij and Muller, 1994, *Geography: Realms, Regions, and Concepts: 7th Edition,* develop essays answering the following questions:

1. Given the complexity of internal issues in Dagestan, predict the future for the people of this republic.

2. One would think that the privatization of agribusiness in Vologda would have forced people to leave the countryside, but the opposite is happening. Why?

3. For both Case Studies, discuss the relationship between access to new markets and relative location. Discuss the possibilities of new markets for each relative to Moscow, to neighboring countries, and to more distant trading partners.

change attitude

▪ PROGRAM 8

Central and Remote Economic Development

LESSONS IN GEOGRAPHY

At the end of Program 8, you should be able to:

1. Describe the spatial representations of a planned economy.
2. Understand the re-emergence of St. Petersburg as an urban center in post-Soviet Russia.
3. Explain how foreign investment and joint ventures are a pivotal part of St. Petersburg's new economy.
4. Evaluate the relationship between housing stock and market pricing in St. Petersburg.
5. Describe the role of central planning in the establishment of Bratsk.
6. List the natural resources required to support hydropower and industry in Bratsk.
7. Assess the role of climate in the past and present settlement of Bratsk.

OVERVIEW

Across the variety of Russian landscapes, geography influences the problems people are facing and the solutions they are considering after the dissolution of the Soviet Union.

In *St. Petersburg: Russia's Window on the West* we see a capital city originally founded to **link** Tsarist Russia to cultured Western Europe. Under communist rule and **isolationism**, the city lost its preeminence, its name, and, in part, the strength of its **economic base** as a **port city**. Emergence of the **market system** has introduced serious problems into its economy, but it has also presented new opportunities for **revitalization**.

Bratsk: The Legacy of Central Planning visits another city built by government decree, although the era in which the city was built and its planned purpose were not the same as St. Petersburg. In the 20th century, Russia's new technologies were powered by the rich **resources** of Siberia. This **exploitation** of Siberia greatly contributed to the Soviet central economy. Russia's use of hydroelectric power and mineral reserves continues to support this city in the midst of an inhospitable region.

Both St. Petersburg and Bratsk were deliberately located and built for specific purposes based upon their particular **locational advantages.** The resulting

functions and **structures** in these cities illustrate the problems the Russian economy faces in the **transition** of each place from communism to a market economy where competition rewards individual initiative.

STUDY RESOURCES

1. Video Program 8: "Central and Remote Economic Development" from *The Power of Place: World Regional Geography*.
2. Unit 3 Map, "Russia's Fracturing Federation" from Latz and Gilbert, 1996, *The Power of Place: World Regional Geography Study Guide*.
3. Chapter 2, "Russia's Fracturing Federation" from de Blij and Muller, 1994, *Geography: Realms, Regions, and Concepts: 7th Edition*.

PREVIEWING QUESTIONS

1. How many time zones are there between St. Petersburg and Bratsk?
2. Locate maritime St. Petersburg and continental Bratsk on a climate distribution map. What are their average temperatures in July and January?
3. How does the latitude of St. Petersburg compare to that of New York, Tokyo, and Beijing.

KEY THEMES

- **Western City**
- **Location, Location, Location**
- **Market Conditions at Home**
- **Planning for Production**
- **Bratsk Braces for the Future**
- **Attachment to Place**

DISCUSSION OF CASE STUDY THEMES

• Western City

St. Petersburg was established in 1703 in the place where the low-lying delta of the Neva River meets the Gulf of Finland. Peter the Great planned this seaport to provide a link to the rest of the European continent. But, in later years, this city that incubated the Communist Revolution was renamed Leningrad and saw its status as Russia's capital lost to Moscow. Now, the physical and social structure of the city is in many ways a living artifact of socialist policies practiced for more than 70 years.

This fourth most populous city west of the Ural Mountains (behind Moscow, London, and Paris) is now open to the influence of western ways and trade in goods from the rest of the European continent. In the city once again called St. Petersburg, reorientation toward the West has begun in the area of free market real estate.

• Location, Location, Location

In a market economy approach to real estate valuation, location is a key factor in setting prices for the conversion of government-subsidized housing to private ownership. Proximity to the historical center of the city, the Central Business District, parks and recreation, and mass transit plays a large role in this new equation.

Not only does the consideration of location and its value occur within the city, but St. Petersburg's international location and its potential comparative advantage adds to market price estimates.

• Market Conditions at Home

The visit to the Pavelko family shows socialism in action. Under Soviet standards, only 96 square feet (9 square meters) of living space were allotted per person. But sturdy, government-backed buildings, shown in the tour of the city, provided each family with its own individual apartment at about 2 percent of the household income. Such guarantees no longer exist for the Pavelkos, whose wages have not risen with the fall of communism. The planned economy took care of citizens, albeit modestly in terms of the "luxury" of space, in a way the new free market does not.

• Planning for Production

As Soviet social structures are dismantled, old policies are often discussed in terms of their failures. Bratsk, however, remains a testament to the power of central planning.

In the 1950s, construction began on a dam for the Angara River. Workers were recruited from areas of the country with a labor surplus and salaries were increased up to 30 percent in order to build the Territorial Production Complex (TPK) and, in later years, to complete the Baykal-Amur Mainline link to the Trans-Siberian Railway. Soviet TPKs were developments of mutually related factories that together used the natural resources, economic resources and infrastructure of a territory.

Everything in Bratsk is the result of one coordinated central planning effort: the dam, power station, railway, aluminum factory — the city itself.

• Bratsk Braces for the Future

In the Soviet period, the big factories of Bratsk supplied basic products for the national economy. Now they are preparing for the international economy. Establishing trade relations and quality standards are not the only new challenges facing Bratsk. Environmental pollution from the TPK, seen in the space shuttle views in the Introductory Programs, endangers the health of the urban population and those living in the vast region beyond the city limits. *complex of factories*

• Attachment to Place

Initially, wages were used as an economic incentive to encourage migration to the inhospitable Eastern Frontier and Siberia. Workers were offered wages up to 30 percent higher than those found in the milder climate of the Russian Core. With a drop off in prior socialist employment rates, one might expect that people are now leaving this area, as is happening in other parts of the Siberian plains.

Even though the climate in the area appears inhospitable, the proud people who have made Bratsk a home remain.

CASE STUDY CONNECTIONS

1. Bratsk and St. Petersburg both bear the imprint of central planning.

2. St. Petersburg is the northern-most major city in the world; Bratsk endures winter temperatures routinely below -22°F (-30°C) and readings of -58°F (-50°C) are not exceptional.

3. Post-Soviet free market conversions are underway in both cities.

4. Citizens of both cities remain in the same housing and work at the same jobs during the conversion.

5. St. Petersburg is a gateway to Europe, and Bratsk is linked to East Asia and the Pacific by rail.

GEOGRAPHERS AT WORK

Olga Pakhonova is a consultant using a Geographic Information System to track the recent phenomenon of real estate pricing in St. Petersburg. The valuation of real estate and land in the West is based on transaction statistics which do not yet exist in Russia. Therefore, she has developed a special database of indicators, such as relative location, to estimate the prices for the 1.5 million apartments to be privatized. Pakhonova's model has been successful in predicting the 300,000 units converted to private ownership so far, providing insight into a host of planning challenges, i.e. transportation infrastructure development, that will confront the city as new centers of economic activity emerge within it.

TEST YOUR UNDERSTANDING

Questions

1. How are the origins of St. Petersburg and Bratsk both similar and different?

2. How does 9 square meters convert to square feet? How many square feet do you live in?

3. How might Bratsk have evolved were it not part of a planned central economy?

4. Why can St. Petersburg be viewed as a window to the West.

Map Exercises

The maps provided at the beginning of this unit can be used to complete this exercise. You may find extra copies of these maps helpful. Also, refer to the maps of Chapter 2, "Russia's Fracturing Federation," from de Blij and Muller, 1994, *Geography: Realms, Regions, and Concepts: 7th Edition*, and any standard atlas to assist in this exercise.

1. Place the following names:

St. Petersburg　　Finland　　　　Gulf of Finland
Sweden　　　　　Norway　　　　Baltic Sea
Estonia　　　　　Russian Plain　　Moscow
Bratsk　　　　　Angara River　　Eastern Frontier
Lena-Tunguska Region　Lake Baykal　Lena River
Baykal-Amur Mainline　Siberia　　　Mongolia

CONSIDER THE DISCIPLINE

CLIMATOLOGY

The **earth's rotation,** coupled with the **differential heating** between equatorial and polar areas, sets up a system of redistribution of warmth from the tropics to the higher latitudes through interaction between the oceans and the atmosphere.

The warm moisture-laden air masses of the Atlantic must cross Western Europe before they reach Russia. As a result, much of the warmth and moisture has been released long before these air masses reach Russia's farmlands. **Low topography** to the north provides no protection for the country from frigid **polar winds.**

The dominance of **extreme climates** in Russia has historically been an important factor in Russia's internal policy decisions as well as its external ambitions. According to de Blij, Russian tsars sought **warm-water ports** because the country's **high-latitude harbors,** like St. Petersburg, froze for several months each year.

Later, Soviet planners diverted whole rivers to bring cultivation to desert lands, and they forcibly settled inhospitable places such as Bratsk for the purpose of **resource extraction.**

To further develop your understanding of this field of physical geography, refer to the Focus on a Systematic Field in Chapter 2, "Russia's Fracturing Federation," from de Blij and Muller, 1994, *Geography: Realms, Regions, and Concepts: 7th Edition.*

CRITICAL VIEWING

Based on the Case Studies in Video Program 8: "Central and Remote Economic Development" from *The Power of Place: World Regional Geography,* and the reading in Chapter 2, "Russia's Fracturing Federalism," from de Blij and Muller, 1994, *Geography: Realms, Regions, and Concepts: 7th Edition,* develop essays answering the following questions:

1. Are the harsh environmental conditions in Bratsk constraints to production or human settlement? Defend your answer.

2. Consider the percentage of your income that pays for your housing. (Is this the same percentage you paid before you became a student? When you finish school, do you plan to change the your housing expense-to-income ratio?) Compare your ratio to the new market pricing system and the planned 2 percent of income paid for rent in St. Petersburg when it was known as Leningrad.

▪ UNIT 4

North America: The Post-Industrial Transformation

NORTH AMERICA:
The Postindustrial
Transformation

| Kilometers | 0 | | 500 | 1000 | 1500 |
| Miles | 0 | | | 500 | 1000 |

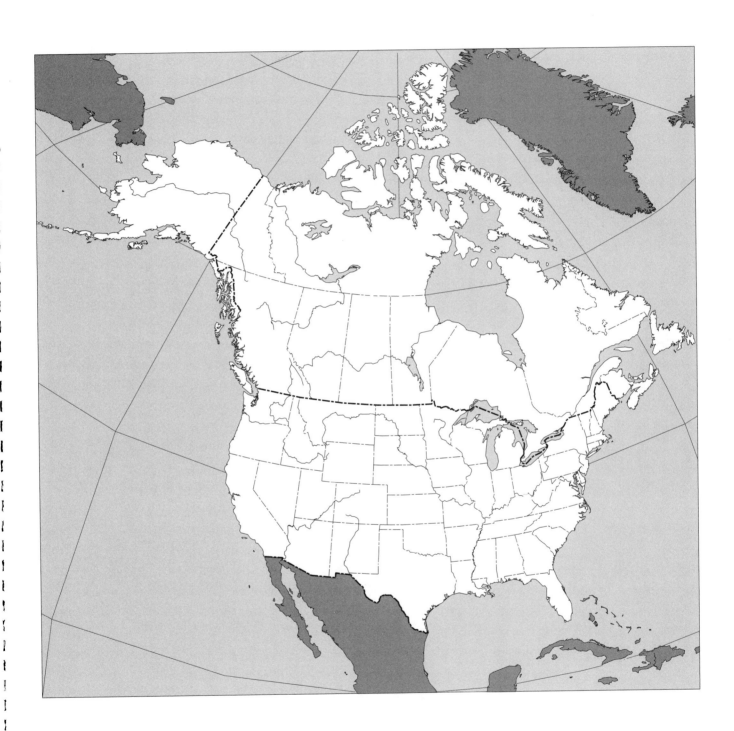

■ PROGRAM 9

Inner vs. 'Edge' Cities

LESSONS IN GEOGRAPHY

At the end of Program 9, you should be able to:

1. Identify some of the processes, patterns and functions of current urban development in the United States.

2. Identify patterns and networks of economic interdependence, as well as the transportation infrastructure, in a North American metropolitan area.

3. Examine how the forces of cooperation and conflict among various ethnic groups in Boston's inner city have influenced the creation of an Empowerment Zone.

4. Discuss how human actions have modified the physical environment surrounding American cities.

5. Analyze the changes that occur in meaning, use, and distribution of resources as they pertain to agricultural and resource lands in the United States.

6. Appreciate how the automobile has uniquely shaped the North American landscape.

OVERVIEW

Unit 4 consists of three programs which address some of the most critical geographical implications of the **postindustrial transformation** of the North American realm.

Program 9: "Inner vs. 'Edge' Cities" looks at Case Studies in Boston, Massachusetts and Chicago, Illinois. *Boston: Ethnic Mosaic* examines issues associated with diverse ethnicity and poverty in the inner city. *Chicago: Farming at the Edge* looks at the pressures of suburban growth on agricultural communities.

The United States and Canada have the most **urbanized** and **mobile** populations in the world. As discussed in the Case Studies, two of the most important results of that mobility are **inner-city abandonment** by the middle class and an increasing loss of **prime agricultural land** to **suburban development**. The land-use pattern that is emerging is one that looks something like a doughnut. This has been created by the middle class fleeing an increasingly poor and empty downtown for life and work in a ring of suburbs and edge cities.

Boston: Ethnic Mosaic visits the home town of some of the nation's finest institutions of higher education and centers for research and development. Boston has a vibrant **Central Business District** (CBD), but it has also experienced middle class flight. The city's **ethnic diversity** is due in part to the attraction of cheap housing for newly arriving immigrants. Inner-city Boston has become, however, a place where tax dollars are simply not adequate to deal with the problems caused by **poverty**.

Inner-city citizens and the city have found that they must turn to the federal government for assistance in solving the resulting social and economic problems.

This Case Study focuses on the **spatial distribution** of some of Boston's ethnic and racial groups and their competition for inclusion in a proposed **Empowerment Zone**. The zone is to include the most impoverished areas in Boston, however, those areas don't necessarily follow established neighborhood boundaries. A geographer is asked to use her professional skills and input from citizens to determine where that proposed boundary will go in the grant application. The stakes are high. A first-place $100-million federal grant could go a long way toward providing job training and social services, and attracting new businesses to Boston's struggling core.

Chicago: Farming on the Edge leaves the downtown core to take a look at the increasing threat that middle class flight is bringing to farmland at the **urban fringe.** No longer are suburban communities dependent upon jobs in the downtown core. The availability of cars and highway infrastructure has enabled a new kind of city to spring up around Chicago. These auto-dependent **edge cities** provide jobs and services to the residents of the surrounding **suburbia,** allowing them to avoid downtown altogether if they choose. Furthermore, these new job centers, already many miles from the old downtown core, enable suburbia to penetrate more deeply into the countryside.

Under these circumstances, enormous pressures are put on farmers to sell their land to developers. In these high-demand areas, fortunes can be made through the sale of increasingly scarce agricultural land.

Together these two Case Studies create a portrait of the urban form that has emerged in postindustrial America. They also reveal some of the serious implications of that form for the people remaining in the urban core, for those people who want to continue their way of life as farmers, and for those who want to preserve some of the nation's best farmland. The forces driving development away from old urban centers are strong. The question is whether those forces should or could be contained.

STUDY RESOURCES

1. Video Program 9: "Inner vs. 'Edge' Cities" from *The Power of Place: World Regional Geography.*

2. Unit 4 Map, "North America: The Postindustrial Transformation" from Latz and Gilbert, 1996, *The Power of Place: World Regional Geography Study Guide.*

3. Chapter 3, "North America: The Postindustrial Transformation" from de Blij and Muller, 1994, *Geography: Realms, Regions, and Concepts: 7th Edition.*

PREVIEWING QUESTIONS

1. Define megalopolis.

2. In your own community, do ethnic minorities tend to live in distinct neighborhoods? Are they spatially segregated?

3. Define metropolitan area. In your metropolitan area, are poor people evenly distributed throughout the area or do they tend to live in distinct places?

4. How would you define a suburban area? What are its characteristics?

5. Approximately how far is your school or workplace from your home? How do you commute (walk, bicycle, public transportation, carpool, single-occupancy automobile)? How many miles is your commute? How long are you on the road?

6. Are there activities that are important to you — such as work, entertainment, shopping, athletic facilities, banking — within either walking or bicycling distance from your home?

KEY THEMES
- **Inner-City Abandonment**
- **Ethnic Immigration**
- **Empowerment Zones**
- **Prime Agricultural Land Near Chicago**
- **Edge Cities**
- **Commuting in North America**

DISCUSSION OF CASE STUDY THEMES

• Inner-City Abandonment

By 1870, immigrants pouring into industrializing American cities sought inexpensive housing close to their factory jobs. Tenements and row houses were filled with as many bodies as it was possible to hold. Conditions were miserable, and, as soon as a family could save enough money, it moved into an apartment in the next best neighborhood, usually one step beyond the center of the city but still within easy commuting distance by public transportation. Tenements would not stay vacant long, as newly arrived immigrants replaced those who were able to move out.

As families became more prosperous, they sought better housing — from tenement, to apartment, to duplex, to single family home usually in the suburbs. This was made possible by the ability of the most affluent families to afford longer commutes and to buy new housing at the suburban edge. As the upwardly mobile vacated their old homes, middle class families would take their place, creating a chain of movement out from the center of the city. This is a pattern that continues to the present.

• Ethnic Immigration

After the United States closed its doors to immigration in the 1920s, industrial employers in need of workers turned to the large unemployed African-American population in the rural south. This recruitment created a new wave of migration into the Manufacturing Belt Cities of the northeast and the midwest. One result of this migration was that, because many whites refused to live with these racially different newcomers, blacks were forced to live in racially segregated neighborhoods. Increasingly, inner cities became places of "color," as whites attempted to flee their own inner-city neighborhoods.

• Empowerment Zones

Today, flight from the inner city is more an issue of class than one of color, as middle class people of all racial and ethnic groups attempt to leave the city for life in the suburbs. This has had a devastating impact economically for nearly all American cities. As inner-city populations become increasingly poor, the need for costly social services becomes critical. Most cities raise the money with which they operate through local property taxes. With the flight of the middle class, property values have decreased, as have property tax revenues. Many cities no longer have the ability to pay for necessary services on their own and must look elsewhere for funding if they are to survive.

The first Case Study follows an Empowerment Zone application for several neighborhoods in South Boston. Such a federal designation could restructure the tax base with the help of $100 million grant. The process is complex because the area to be defined for the application includes a number of racially and ethnically separated neighborhoods, each with its own needs. Newly configured boundaries are created for the Empowerment Zone to describe the areas more precisely than do traditional neighborhood boundaries.

As seen in the video, news of the second-place $25 million grant was positively received by the community leaders of the most impoverished neighborhoods in the city.

• Prime Agricultural Land Near Chicago

Cities throughout the world have historically been located near water. They tended to be built on relatively flat land where it is less costly to build. These places are also very often where some of the best soils are found, as is the case of the agricultural heartland of the United States.

As long as cities remained relatively small and compact, there was little concern about the permanent loss of farmland to real estate development. However, as urban areas, such as Chicago, have gobbled up farmland that is 60 miles or more from the urban core, many people have become concerned that a shrinking base of farmlands and sprawling congestion may outweigh the benefits of further expansive growth.

• Edge Cities

The geography of America's past is undergoing a postindustrial transformation in McHenry County, Illinois. Along Interstate 90 heading west from Chicago, new business subcenters such as Des Plaines, Schaumburg and Marengo are home to more and more residents. First, shopping malls sprouted up in these once purely residential areas, then business followed their employees to this suburban fringe.

New suburban developments and 'edge' cities have prospered, often at the expense of aging inner-city cores as well as surrounding farmlands. The increasing exodus of the middle class from inner cities has created pockets of intense poverty at city centers. Furthermore, the cycle of development of suburbia and its supporting 'edge' cities has allowed for an even greater penetration of urbanization into the countryside.

• Commuting in North America

North America has the most highly urbanized population in the world; it is also the most mobile. Vast networks of superhighways, commuter airplanes and railroads connect this realm's cities from coast to coast, distinguishing this realm from Europe where there is a greater balance between mass and personal transit modes. But it is the mobility of the daily commuter — a driver who typically considers commuting in terms of time not distance — that has most dramatically expanded the McHenry suburbia more than two hours out from Chicago's CBD.

CASE STUDY CONNECTIONS

1. Both inner-city abandonment and suburban expansion are shaped by transportation infrastructure.

2. Both farm life and inner-city life are threatened by middle-class urban flight.

3. Just as mobility has increased the choices for the middle class, it has limited the choices for many living in the inner city who are trapped in a cycle of seemingly inescapable poverty.

4. The subdivision of land into small parcels — whether at the 5-acre level of exurbia, the quarter- to half-acre lot of the standard suburban neighborhood, or the intense development of CBDs — renders land inadequate for farming.

5. Government intervention may be necessary to protect farmlands, as well as inner-city communities.

6. A geographic information system is used in both Case Studies to analyze the spatial impact of land-use conversion.

GEOGRAPHERS AT WORK

Geographers study spatial distributions of all kinds across the earth. In Program 9, we use a geographic perspective to analyze the patterns of urban settlement.

In *Boston: Ethnic Mosaic*, a geographer is asked to create a map of places in the city of Boston that will become, if funded, an Empowerment Zone. Both social and technical issues have to be addressed. Linda Haar uses a geographic information system — which allows her to combine or intersect demographic, economic and spatial information — and citizens' comments to create an Empowerment Zone boundary that best reflects community needs and expectations. The geographer's ability think in spatial terms allows her to integrate the social, economic and spatial aspects of poverty in inner-city Boston.

In *Chicago: Farming on the Edge*, another geographer, Richard Greene of Northern Illinois University, uses his skills to develop new knowledge about his community. Here, too, through the use of a geographic information system he is able to analyze and understand where suburban development can have the most harmful effects. By combining information on soils, hydrology, and other data, he is able to see in spatial terms which farmland is most valuable for agriculture and where development is most likely to endanger that land.

TEST YOUR UNDERSTANDING

Questions

1. What is meant by an 'edge' city? How does it differ from the central city in terms of economic activity, who lives or works there, and what it looks like? What are its strengths and weaknesses? What are the strengths and weaknesses of the central city?

2. What is postindustrial transformation?

3. Why do you think a geographer was chosen to work on the Empowerment Zone grant application for Boston's inner-city neighborhoods?

4. Why is farmland permanently lost when it is developed? How might development impact soils? Why do you think many people are concerned about the loss of farmland in the United States today?

Map Exercises

The maps provided at the beginning of this unit can be used to complete this exercise. You may find extra copies of these maps helpful. Also, refer to the maps of Chapter 3, "North America: The Postindustrial Transformation," from de Blij and Muller, 1994, *Geography: Realms, Regions, and Concepts: 7th Edition*, and any standard atlas to assist in this exercise.

1. Place the following names:

Boston	Massachusetts	Washington, D.C.
New York City	New York State	Northeastern U.S. Megalopolis
Chicago	Illinois	Lake Michigan
Mississippi River	Marengo	Des Plaines

2. On a metropolitan area map, locate and map (shade in) the agricultural land in your area. What kind of agriculture takes place there? If you have been to these areas, have you seen "for sale" signs or new developments? Is there a pattern to their locations? Based on where the edge of development is now in your area and the existing transportation links (highways, roads, rail), where do you think farmers can get the most money for their land?

CONSIDER THE DISCIPLINE

THE ROLE OF TRANSPORTATION IN URBAN GEOGRAPHY

Where people live and work has historically shaped **settlements.** And where they are able to live and work is largely determined by the available **transportation modes** and **infrastructure**. The way cities have grown can be organized into "eras of **intra-urban** structural evolution" based on transportation as described by John Adams in de Blij and Muller's Chapter 3, "North America: The Postindustrial Transformation," in *Geography: Realms, Regions, and Concepts: 7th Edition.*

Prior to 1888 when most people could only get around on foot, settlements were much smaller and more compact. During this **Walking-Horsecar Era** everything — housing, shopping, government, banking, entertainment, medical care — needed to be within walking distance.

After the invention of the electric traction motor, a 30-minute commute stretched the edges of the city. During this new **Electric Streetcar Era** (1888-1920), development followed trolleys radiating from the center of the city. As people moved to suburbs, land-use specialized at the city center, creating a distinction between activities occurring in the CBD and the city's other zones.

The city stretched farther with the advent of the automobile. The car allowed for unprecedented individual mobility. The **Recreational Automobile Era** (1920-1945) began the massive suburbanization of the American population. In 30 minutes people could commute beyond the range of the streetcar, and development was no longer constrained along rail lines.

The **Freeway Era** (1945 to the present) saw the full impact of the automobile. In the 1950s, safe high-speed automobile travel became possible on America's new Interstate Highways. Commuters pushed suburbia further from CBDs, often into prime agricultural land. Many freeways were designed to go directly into city cores for easy commuting. These massive structures required the wholesale destruction of many inner-city neighborhoods.

The Freeway Era prompted more people to live in suburbs rather than in cities. Today, the suburban population is great enough to encourage employers to move their operations to the suburban fringe, where land is cheap and potential workers are plentiful. Radiating arterials, complemented by a series of **beltways,** ring cities, allowing for **inter-suburban** commuting and creating a doughnut pattern of transportation which ignores the CBDs altogether.

This infrastructure has made the emergence of **suburban downtowns** possible. These **auto-dependent** employment-recreation-shopping centers have sprung up at freeway intersections around most American cities, forming what de Blij refers to in the video commentary as the **pepperoni pizza urban growth model**.

To further develop your understanding of this discussion, refer to Focus on a Systematic Field in Chapter 3, "North America: The Postindustrial Transformation," from de Blij and Muller, 1994, *Geography: Realms, Regions, and Concepts: 7th Edition*.

CRITICAL VIEWING

Based on the Case Studies in Video Program 9: "Inner vs. 'Edge' Cities" from *The Power of Place: World Regional Geography,* and the reading in Chapter 3, "North America: The Postindustrial Transformation," from de Blij and Muller, 1994, *Geography: Realms, Regions, and Concepts: 7th Edition*, develop essays answering the following questions:

1. Based on the Case Studies, is there a relationship between the plight of the inner-city poor and the people living in the suburbs? Defend your answer.

2. Some communities use a mechanism called an "urban growth boundary" to determine how far urban/suburban development can extend from an urban core in a given period of time, for example, 20 years. Using a map of your metropolitan area, draw a proposed urban growth boundary.

   • What data would you need to have in order to draw such a boundary?

   • How would you determine how much developable land to include?

   • Would you limit the number of people in your area or allow for population growth through increasing density? How?

   • Would you limit the lot size for each house?

   • If you choose to accommodate more residents within the boundary, how would you do this in your model?

   • What other approaches might be used to control urban growth while preserving farmlands?

PROGRAM 10
Ethnic Fragmentation in Canada

LESSONS IN GEOGRAPHY

At the end of Program 10, you should be able to:

1. Explain how the forces of cooperation and conflict among people influence the geopolitical boundaries of Québec.

2. Describe how culture and experience influence residents' perceptions of Vancouver.

3. Briefly discuss the pattern and composition of immigration to Québec.

4. Recognize the pervasive influence of the United States on its Canadian neighbor and Canadians' reaction to this influence.

5. Understand that the western margins of Canada are becoming increasingly oriented toward the Pacific Basin.

OVERVIEW

Québec: An Island of French examines the **French-speaking** population within Montréal, its turbulent history with the **English-speaking minority**, and the ongoing efforts to resist **linguistic domination** in North America. The Case Study focuses in particular on the city's large **immigrant population** and its importance in the efforts of francophones to maintain their **majority** status.

Vancouver: Hong Kong East focuses on this emerging **Pacific Rim** metropolis and the consequences of a recent **influx** of wealthy Hong Kong Chinese immigrants into the area. Older, well-established "Anglo" neighborhoods fight to preserve their **cultural landscape**, as new Asian residents tear down older homes to erect larger, more modern dwellings whose styles are radically different.

STUDY RESOURCES

1. Video Program 10: "Ethnic Fragmentation in Canada" from *The Power of Place: World Regional Geography*.

2. Unit 4 Map, "North America: The Postindustrial Transformation" from Latz and Gilbert, 1996, *The Power of Place: World Regional Geography Study Guide*.

3. Chapter 3, "North America: The Postindustrial Transformation" from de Blij and Muller, 1994, *Geography: Realms, Regions, and Concepts: 7th Edition*.

PREVIEWING QUESTIONS

1. Locate Montréal and Vancouver on a world map. How far apart are they in miles and kilometers? What are the similarities in their relative location to the United States?

2. In which Canadian provinces are Montréal and Vancouver located?

3. What is the predominant language used in Montréal? In Vancouver?

KEY THEMES

- **The Struggle to Preserve French Culture in a Sea of English**
- **The Demographics of Linguistic Domination**
- **Immigrant Assimilation: French or English?**
- **The Hong Kong Diaspora**
- **Urban Geography of Vancouver**
- **The Global Meets the Local**

DISCUSSION OF CASE STUDY THEMES

• The Struggle to Preserve French Culture in a Sea of English

Nearly half of Québec's 7 million people live in the city of Montréal. It is here, surrounded not only by English-speaking Canada, but by the largest and most influential English-speaking country in the realm, the United States, that the battle between French and English is most heated. Québec's French-speaking majority, the Québécois, recognize that protecting the language is the single most effective means of preserving culture. And protection from English is just what the Québec government is seeking.

Historically, Montréal's English-speaking minority held the economic power in the province. Although English speakers represented only one-quarter of the population of the city, the language used in business and commerce, and on many public signs, was English. Fearing the decay of francophone culture, the threatened French-speaking majority took a number of calculated steps in the 1970s aimed at securing priority status for their language and culture. An example of this was enacting legislation which banned unilingual English signs. Between the early 1970s and 1980s, 20 percent of English speakers and their companies left the province.

• The Demographics of Linguistic Domination

Although small pockets of francophones exist throughout North America, by far the largest population resides in Québec. Like most of North America, Québec's birthrate began to decline in the 1970s as it transformed to a postindustrial economy. But the Québécois suffer from the lowest birthrate in all of Canada. The situation is considered so serious that the provincial government pays families $6,000 Canadian dollars for every child born after the first two; but still the birthrate is low.

• Immigrant Assimilation: French or English?

To compensate for this low birthrate, Québec has looked outside its borders and welcomed immigrants from around the world. This rapidly growing immigrant population poses a great challenge to the Québécois. An immigration office campaign slogan trumpets the French language as the key to success in Québec, and to assure that newcomers agree, the government pays them. Immigrant families can receive up to $200 Canadian dollars a week to be taught by the government how to assimilate into French-speaking society.

But many immigrants do not feel obligated to help preserve French language and culture. Despite government inducements and laws such as English signage restrictions, the immigrants believe that English is the key to success for their children. Yet in Québec, choosing English has a price. Although English public schools do exist in Montréal, the only free education available by law to immigrants is a French one.

Immigrants are arriving and multiplying as such a rapid rate that some observers estimate the Québécois could make up less than half of the population of Montréal by 2001.

• The Hong Kong Diaspora

The cultural landscape of Vancouver, B.C., is being transformed by the influences of Asia. Nine thousand immigrants per year come to Vancouver from Hong Kong, where wealthy families fear the 1997 return of this British protectorate to Communist Chinese rule. The Hong Kong diaspora has found a safe haven that is relatively close to home in Vancouver. Immigration across the Pacific Rim is not new. Vancouver has an old and well-established Chinatown. But most of today's immigrants are not moving to this part of Vancouver.

• Urban Geography of Vancouver

Vancouver's east side has always been home to non-English-speaking immigrants; some were from Europe, some were from Asia, but all were working class people, and their smaller houses still dominate the landscape. Chinatown is located in this more densely populated east side.

On the west side, greater wealth is reflected in larger lots and an abundant urban forest. It is this greener, more affluent part of the city in which the new Asian elite has chosen to live.

• The Global Meets the Local

In the past, homes on the west side often reflected Anglo traditions, with Tudor styling or large gardens. But new homes are being built on the west side which reflect the tastes of new Asian buyers. As developers rush to accommodate affluent new arrivals from Hong Kong, hundreds of old homes have been torn down to make way for spacious, more modern designs. Not only are the remaining long-time residents of western Vancouver worried about the clash of styles between the old and the new, they also fear the changing culture reflected by the grand nature of these custom-built homes. Yet these new cultural influences, accepted or not, are a new reality for Vancouver. All along the Pacific Rim, the old diaspora from Europe is meeting the new diaspora from Asia. These global cultures are meeting on a very local and personal level in the high-income neighborhoods of Vancouver.

CASE STUDY CONNECTIONS

1. Canada is a nation of immigrants, but the homelands of these newcomers differ dramatically when one compares British Columbia and Québec.

2. Canada's external orientation differs dramatically from east to west; eastern Canada is culturally and economically tied to the Atlantic world; western Canada, to the Pacific world.

3. Promotion of a multi-ethnic society challenges the future of Canadian national unity.

4. Global forces of trade and migration are transforming the communities and neighborhoods of Montréal and Vancouver.

GEOGRAPHERS AT WORK

David Ley is an urban geographer at the University of British Columbia, in Vancouver, Canada. He is widely regarded as one of the leaders in the field of urban geography, focusing on the social organization, internal structure, and shape of cities.

In *Vancouver: Hong Kong East,* he comments cogently on the conflicts that have evolved at the neighborhood level as Chinese immigrants' values clash with those of Anglo residents over the type and style of newly constructed housing. His use of the geographic concept of scale to explain his observations is rich with meaning: on the one hand, the very scale of the newer houses is at odds with traditional Tudor homes; on the other hand, the economic and political forces that encourage landed-immigrant status for Chinese Canadians are national and global in scope.

In the final analysis, differing views of the purpose and function of houses in Vancouver are at the root of the conflict. Ultimately, the neighborhood undergoing change will mediate the matter through revision of the area's zoning code. In the course of ameliorating heated emotions, geographic insight reminds us that the rules and regulations governing how land is used, and houses built, is a reflection of a variety of urban community values that must be blended together.

TEST YOUR UNDERSTANDING

Questions

1. Describe the measures taken by the Québec government to promote its French national identity.

2. How willing are immigrants to Québec to embrace French culture given that the majority of Canadians are English speakers and that Canada borders the United States?

3. Account for the Pacific orientation of Vancouver, British Columbia in terms of immigration policies and trade patterns.

4. Why is Vancouver's urban geography a factor in explaining ethnic conflict in this city?

Map Exercises

The maps provided at the beginning of this unit can be used to complete this exercise. You may find extra copies of these maps helpful. Also, refer to the maps of Chapter 3, "North America: The Postindustrial Transformation," from de Blij and Muller, 1994, *Geography: Realms, Regions, and Concepts: 7th Edition*, and any standard atlas to assist in this exercise.

1. Place the following names:

Québec	Montréal	St. Lawrence River
New York City	Atlantic Ocean	Chicago
Ontario	Ottawa	Pacific Ocean
British Columbia	Vancouver	Columbia River
Vancouver Island	Puget Sound	Seattle
Hong Kong	Taiwan	Tokyo

CONSIDER THE DISCIPLINE

INTERNATIONAL TRADE COMMUNITIES

Supranationalism, the voluntary grouping of countries for the purpose of promoting trade and political arrangements supportive of regional development, is one of the most intriguing phenomena of the late 20th century. The **North American Free Trade Agreement** (NAFTA) is one such effort to promote the integration of the economies of Canada, the United States, Mexico, and perhaps in the future, Chile.

NAFTA, ratified by each country by the mid-1990s, seeks to reinforce the continental integration of the economies of these three countries. It creates the world's largest trading bloc estimated to be valued at over $6.5 trillion dollars with nearly 400 million consumers. **Economic integration** is promoted primarily through a reduction of tariffs, quotas, and other impediments to the flow of goods, services, and capital across the borders separating each country.

NAFTA is not the only effort of this type now occurring on the world stage. APEC, the **Asia Pacific Economic Cooperation forum**, a second important example of regional cooperation, also attempts to facilitate economic liberalization. This forum, which consists of 18 economies on both sides of the Pacific, though centered on the Western Pacific Rim, commands 40 percent of the world's population, 40 percent of the world's exports, and a staggering 50 percent of its Gross Domestic Product. In the case of the U.S., 62 percent of its exports are destined for APEC countries.

The difficulty APEC faces — finding common purpose in so large and diverse a geographic area — is daunting when compared to the region comprising NAFTA. Contained within the APEC region are the major world economies of Japan and the U.S., the communist giant China, authoritarian and Islamic Indonesia, and multi-ethnic Malaysia. The reality of defining common identity must be balanced against economic accomplishments and potential; one is reminded repeatedly, and with good reason, that APEC is not comparable to the cultural and political cohesion — albeit contentious and diverse — of either North America or Europe.

Even so, the emergence of APEC, a collection, if not a community, of common interests across the Pacific, is remarkable and important. It parallels more formal supranational governmental efforts through NAFTA as well as the even more ambitious idea of integrating the political and economic diversity of Europe through a European Union (EU). The relationships between these three world trade blocs should not be seen as inevitably competing with one another. The concerted effort by the Canadian government to encourage Hong Kong Chinese nationals to immigrate to Vancouver, British Columbia suggests that both a north-south as well as an east-west axis of trade and interchange of ideas have become centered on the western margins of North America. In the years ahead, the intersection of these two axes may serve to distinguish this region within an increasingly integrated world economy.

To further develop your understanding of supranationalism and international trade, refer to the box "North American Free Trade Agreement" in Chapter 3, "North America: The Postindustrial Transformation," from de Blij and Muller, 1994, *Geography: Realms, Regions, and Concepts: 7th Edition*.

CRITICAL VIEWING

Based on the Case Studies in Video Program 10: "Ethnic Fragmentation in Canada" from *The Power of Place: World Regional Geography*, and the reading in Chapter 3, "North America: The Postindustrial Transformation," from de Blij and Muller, 1994, *Geography: Realms, Regions, and Concepts: 7th Edition*, develop essays answering the following questions:

1. Compare and contrast the situations in Québec and Vancouver to your local area. Has there been an influx of a minority population? Is there more than one language spoken? Would your area be considered diverse or homogeneous?

2. Have building trends in your area changed? Have styles tended to deviate depending on the individual builder/owner? Describe how the changes in housing styles in Vancouver are perceived as a loss of cultural identity by the Anglo residents.

3. Considering NAFTA as background, it would seem that the concerted effort by the Canadian government to encourage Hong Kong Chinese nationals to immigrate to Vancouver, B.C. leads to a reorientation of the flow of trade and the interchange of ideas along the Pacific Rim. Discuss this intersection and its influence in an increasingly integrated world economy.

▪ PROGRAM 11

Regions and Economies

LESSONS IN GEOGRAPHY

At the end of Program 11, you should be able to:

1. Describe the factors which contribute to the semi-arid region of eastern Oregon becoming a site for large-scale agriculture.

2. Examine the forces of cooperation and conflict that are at play in water resource management issues in eastern Oregon.

3. Discuss how humans have modified the physical environment of the Marginal Interior of the United States.

4. Explain why the Continental Core of the U.S. is dominant in automobile production based on its locational advantage.

5. Compare and contrast Japanese lean production methods to the American tradition of mass production.

6. Describe how automobile manufacturing has changed the landscape of the U.S. Midwest and explain why the Continental Core is an ideal site for automobile factories.

OVERVIEW

Oregon: The Fight for Water looks at agricultural production in the **regional transition zone** from the **West Coast** to the **Marginal Interior** of eastern Oregon. In this region, technology has enabled people to harness scarce **water resources** to support agricultural production, but at an **environmental cost.** We meet farmers who rely upon the Columbia River for **irrigation** for their fields and for **transportation** of their products — 30 percent of which are shipped across the Pacific to Asia. A local Native American tribe is demanding that the diversion of the river be limited in order to revive the dwindling salmon runs that rely on the Columbia and its tributaries to spawn.

    U.S. Midwest: Spatial Innovations focuses on the geographic distribution of Japanese auto plants throughout the midwestern United States, and explores the **spatial nature** of Japanese **just-in-time** production techniques. Proximity to the majority of U.S. consumers and the sophistication of the Midwest's **transportation infrastructure** are key reasons for Japan's decision to locate in this region. The Case Study also explores the U.S. automotive industry, its history of **competition** with Japan, and its gradual incorporation of Japanese production methods to meet the high standards set by its competitor. Competition and cooperation among industries and nationalities have led to new levels of productivity and quality in this area of the **Continental Core.**

STUDY RESOURCES

1. Video Program 11: "Regions and Economies" from *The Power of Place: World Regional Geography*.
2. Unit 4 Map, "North America: The Postindustrial Transformation" from Latz and Gilbert, 1996, *The Power of Place: World Regional Geography Study Guide*.
3. Chapter 3, "North America: The Postindustrial Transformation" from de Blij and Muller, 1994, *Geography: Realms, Regions, and Concepts: 7th Edition*.

PREVIEWING ACTIVITIES

The maps provided at the beginning of this unit can be used to complete this previewing activity. Also, refer to any standard atlas or Chapter 3, "North America: The Postindustrial Transformation," from de Blij and Muller, 1994, *Geography: Realms, Regions, and Concepts: 7th Edition*, to assist in this previewing activity.

1. Locate the Marginal Interior and the Continental Core on a map of the U.S.
2. Confirm the location and the extent of the Columbia/Snake River watershed.
3. Examine the population densities of the Marginal Interior and the Continental Core on a population distribution map.
4. Compare the differences of each region on a physiographic map. Do the same with a transportation map.

KEY THEMES

*   **Rain Shadow Precipitates a Regional Transition Zone**
*   **Transformation of the Marginal Interior**
*   **Water Fuels an Economy at Home and Abroad**
*   **Deserts and Dams — But at What Cost?**
*   **Salmon and the Windward Side of the Columbia Watershed**
*   **Economic Geography in the Midwest: Diffusion of Innovation**
*   **Lean Production — Supplies Delivered Just in Time**
*   **Distance and Accessibility — Transportation Networks**
*   **Globalization in the Continental Core**

DISCUSSION OF CASE STUDY THEMES

• Rain Shadow Precipitates a Regional Transition Zone

Water resource management issues in Oregon are framed by the state's location astride a regional transition zone from the West Coast to the Marginal Interior.

As moist maritime air moves inland from the Pacific Ocean, the Cascade Range thrusts the air upward, causing it to cool. This increases the relative humidity of the air, forming clouds and altitude-induced precipitation known as *orographic* (mountain) rainfall. The result is a heavily vegetated, moist environment on the Cascade's *windward*, or exposed, side which faces the winds flowing across it. As air descends on the protected, or *leeward,* side of the Cascades, it warms, and the relative humidity decreases. The resulting rain shadow effect creates a semi-arid environment in eastern Oregon.

• Transformation of the Marginal Interior

No realm on earth has been shaped by human technology the way North America has. This transformation is reflected on the western landscape of the U.S. by dams and water projects. The U.S. Bureau of Reclamation has built over 700 dams — most in a region Professor H.J. de Blij calls the Marginal Interior. One such water project is located on the Umatilla River in eastern Oregon. Now, as water flows down the Umatilla, instead of continuing along its original course, a government-subsidized dam diverts water via canals to dozens of farms. As seen in the satellite photos, hundreds of fields are now irrigated in circular patterns near the junction of the Umatilla and the larger Columbia River.

• Water Fuels an Economy at Home and Abroad

For much of the past half-century, the U.S. government has played a major role in harnessing water resources for landowners in the Marginal Interior in an effort to support a productive agricultural economy. Today, eastern Oregon farmer Chet Pryor depends on center-pivot irrigation to grow carrots, wheat, alfalfa, and potatoes — over 15,000 tons of potatoes a year not only for the domestic market but for export as well. With the benefits of a government-subsidized water supply, Pryor and his fellow farmers in Umatilla Valley produce almost $100 million of agricultural crops a year. Oregon's agricultural exports are a bright spot not only in the state's but also in the nation's balance of trade.

Farmers in eastern Oregon use water not only for irrigation, but for transportation. Most of the french fry potato harvest and other crops produced along the Umatilla begin their journey to market on barges that travel west on the Columbia toward the Pacific. In the french fry example, 30 percent of the local harvest then continues across the ocean to the Western Pacific Rim. One reason barges can navigate the once-rocky and fast-moving Columbia is the presence of multipurpose dams. From its headwaters in British Columbia to its mouth at the Pacific Ocean, 14 dams along the Columbia now raise the water level to navigable depths for most of its course.

Water resource management projects to meet irrigation needs, provide inexpensive energy through hydroelectric power generation, increase navigability, and regulate flood waters have all changed the nature of the Columbia River in order to fuel the agricultural and urban sectors of the state.

• Deserts and Dams — But at What Cost?

Efforts to develop Oregon's economy through water management projects have severely affected the life cycle of salmon as they migrate from river to ocean and back again, thus wreaking havoc on those native communities once dependent on abundant salmon harvests.

Before the U.S. Army Corps of Engineers was charged with increasing the navigability of the Columbia River early this century, millions of salmon entered the river each year, travelling upstream and spawning in tributaries like the Umatilla. Traditionally, native tribes depended on fish for their survival. Now, under

U.S. Bureau of Reclamation regulation of this tributary, the Umatilla has been diverted for irrigation and the low flow of the river can support neither thriving salmon runs or the native culture based on their once-abundant harvest.

Because water is scarce in this region, the Bureau of Reclamation initially intended that the water from their developments go only to a limited number of farmers. But some of those authorized as irrigation districts conserved their free allocations of this valuable commodity and sold it at a profit to other Umatilla Valley farmers like Pryor.

Desperate to save their salmon, the Indians are threatening legal action based on long-standing treaty rights with the federal government to stop such "water spreading" and preserve this resource for their needs. Even though the unauthorized distribution of the irrigation districts' conserved water was most likely illegal, the tribes are not sure they want to shut off the water supply to farmers like Pryor. The farmers in this area have accomplished just what the government, through the Bureau of Reclamation, intended — they made the desert bloom. Now, reducing the supply of irrigation water would hurt not only the farmers, but the whole region.

• Salmon and the Windward Side of the Columbia Watershed

For millions of years, salmon laid their eggs in small rivers and tributaries like the Umatilla. The young smolts were then washed down the unrestricted Columbia to mature in the Pacific. Years later, the adult salmon returned upstream to spawn in their birthplace. Enormous multipurpose dams on the Columbia now impede both legs of this journey. Those salmon that survive the trip downstream over the dams can now return upstream only with the help of fish ladders — steps of running water built to bypass the dams. Taking into account the impact of other human interference on salmon tributary spawning grounds, such as low in-stream flow due to irrigation diversion projects, biologists point to these dams on the main stem of the Columbia as the main cause of dramatically declining fish counts. Both biologists and engineers have proposed solutions, but most are expensive and would mean less water for vital economic activities ranging from transportation to inexpensive hydroelectric power generation.

Most of the electricity in the region is generated by Columbia River dams in the Marginal Interior. But transmission lines extend across the regional transition zone to consumers on the West Coast, where the state's population and ultimately political power are concentrated.

Residents in both regions benefit from water policies pursued in the dry interior. Not only do the dams provide cheap electricity, but many West Coast jobs are tied to trading and shipping the agricultural products and lumber barged downstream from eastern Oregon.

As the population along Oregon's coast has increased, residents have become more protective of the environment in which salmon once symbolized both wilderness and prosperity. Faced with declining catches, ocean-going commercial fishermen have joined West Coast environmentalists and leeward Indian tribes across this regional transition zone in calling for solutions to the conflict between the management of scarce water resources and salmon preservation.

• Economic Geography in the Midwest: Diffusion of Innovation

In *U.S. Midwest: Spatial Innovations*, an analysis of the changing economic geography of the United States is presented by James Rubenstein of Miami University of Ohio.

Economic geographers like Rubenstein are interested in the reasons for the location, concentration, and distribution of economic activity. Through the identification of land-use and landscape patterns reflecting selected types of economic activity, geographers can explain why particular places grow or decline over time, and make predictions about the future.

The questions that drive Rubenstein's research are threefold: he is interested in where major manufacturers locate, the relative location of new suppliers of parts for these large operations, and the diffusion of Japanese production techniques in the U.S. automobile manufacturing industry.

The principle of diffusion can be described as the geographical spread of an idea from a given point over an increasingly wider area. Geographers study the spreading of an idea or innovation from its source outward across the landscape at varying scales: local, national, and international. In the case of the U.S. automobile industry, research often focuses on the factors that promote and impede adopting the Toyota Production System, or lean production method, a highly efficient production method that has revolutionized the automobile assembly process.

• Lean Production – Supplies Delivered Just in Time

Lean production combines teamwork, bottom-up engineering, worker involvement, just-in-time delivery and the philosophy of continuous improvement, all to great advantage. One aspect of lean production is particularly spatial — just-in-time delivery.

Just-in-time is an inventory control system based on the idea that instead of maintaining large inventories, each workstation on an assembly line keeps only a few hours supply of parts on hand at any given time. Very small lots are ordered by workers as parts are used. Once an installer finishes with one container of parts, a new one drops into place, and the worker pulls and processes its *kanban*, that is, its inventory ticket. The kanban is sent to nearby parts suppliers where parts are packaged and delivered to the factory for installation within about eight hours. No paperwork, no middle managers, and most importantly, no space-consuming idle inventory.

As just-in-time manufacturing is reshaping the factory floor, on a much larger scale it is changing the geography of parts supply networks. The old mass production process more easily accommodated parts shipped from around the world — an engine from Germany, a transmission from Japan, bumpers from Korea — as an integrated individual car was produced. But just-in-time puts a premium on proximity and quality control. Parts suppliers in the Midwest have settled along two major highways, now known as the Kanban Highways, near the assembly plants they serve. The new geography of parts supply, along with the rest of the lean production system, has been grafted onto the North American landscape by Japanese transplants.

• Distance and Accessibility – Transportation Networks

Why did Japanese auto makers set up production in the Midwest? The answer is distance and accessibility. The major auto makers of Japan have virtually eliminated the greatest portion of their shipping costs by setting up shop in the midst of their market. The centrality of the Midwest, not only to the large consumer base, but also to existing parts manufacturers and suppliers, was a key element in the transmigration of this industry.

The placement of these new factories was linked to transportation networks already existing in the U.S. infrastructure. When choosing factory sites, the Japanese considered the spatial proximity of existing parts manufacturers to new factory sites and to defined highway systems. This increased the efficiency of parts ordering while, simultaneously, the just-in-time production method decreased the need for large inventory storage and error reprocess areas within production plants.

• Globalization in the Continental Core

The influx of Japanese automobile manufacturers into the region Professor H.J. de Blij labels the Continental Core represents the spatial dimensions of an open-door economic policy that has allowed foreign investors to open and maintain factories and businesses in the U.S. The cost associated with bringing finished products to consumers was a major factor in the decision to locate Japanese plants at sites within 500 miles of the majority of the U.S. population. Not only have Japanese car makers reaped benefits from their U.S.-based ventures, these ventures have also boosted local economies through increasing employment opportunities in the Midwest.

The U.S. automobile industry has benefited from the globalization of Japanese automobile manufacturing. Cooperation between the Japanese and the Americans has allowed for the transfer of key Japanese production techniques to U.S. industries. While the transfer of ideas appears to be in one direction, it is actually reciprocal when one considers the history of the automobile production process. Japanese insights represent an updating of production techniques pioneered earlier in the century, the so-called Ford Model, as well as quality control concepts originally conceived in the U.S. some 40 years.

CASE STUDY CONNECTIONS

1. Although the United States is undergoing a postindustrial transformation, it continues to have a strong agricultural sector and to emphasize maintainence of a strong manufacturing center.

2. Population density in the Continental Core is much higher than in the Marginal Interior. This difference is related to the type of economy pursued in each region.

3. International trade and investment are transforming the economic geographies of both the Continental Core and the Marginal Interior.

4. Transportation networks are very important to both the Marginal Interior and the Continental Core, as seen in the development of dams to create navigable rivers in Oregon and the proximity of manufacturing plants to major highways in the Midwest.

5. Technological innovations are helping both automobile manufacturers and those employed in the agricultural sector improve their yield and productivity.

6. The problems of both preserving salmon runs and producing higher quality automobiles are being addressed by the adoption of new ways of thinking and the diffusion of new ideas through cooperation of diverse cultural groups.

GEOGRAPHERS AT WORK

Don Schjeldahl, a specialist in location theory, is a site selection consultant at The Austin Company.

Knowledge of geography is essential to Schjeldahl's applied research; it allows him to synthesize data about site location, ranging from the physical characteristics of a site to its economic position relative to the needs of the automobile manufacturing investors with whom he consults. Of all the factors considered in facility placement of Japanese automobile assembly plants in the 1980s, Schjeldahl's research concluded that travel time from parts suppliers was most important. He recommended that his clients position their assembly plants at the "center" of a circular functional area within which travel time from parts suppliers was no more than one to two hours.

TEST YOUR UNDERSTANDING

Questions

1. How have dams for irrigation, and the multipurpose dams developed for navigation and hydropower, changed the salmon habitat in Oregon.

2. How do land-use and population patterns differ on the West Coast and in the Marginal Interior.

3. Explain the geographical concepts of orographic effect and rain shadow.

4. Discuss the economic reasons for the decision of Japanese automobile manufacturers to locate in the U.S. Midwest.

5. How does the just-in-time production method work?

6. Discuss the process of globalization of the U.S. automobile industry.

Map Exercises

The maps provided at the beginning of this unit can be used to complete this exercise. You may find extra copies of these maps helpful. Also, refer to the maps of Chapter 3, "North America: The Postindustrial Transformation," from de Blij and Muller, 1994, *Geography: Realms, Regions, and Concepts: 7th Edition,* and any standard atlas to assist in this exercise.

1. Place the following names:

West Coast	Continental Core	Marginal Interior
Oregon	Columbia River	Snake River
Umatilla River	Portland	Cascade Mountain Range
Washington	Seattle	Pacific Ocean
Indiana	Ft. Wayne	Illinois
Michigan	Detroit	Kentucky
Ohio	Dayton	Tennessee
St. Louis	Ohio River	Mississippi River

CONSIDER THE DISCIPLINE

THE RAIN SHADOW EFFECT

A **natural resource** can be described as a substance which forms part of the earth's natural composition and is of value to people. Wise management of the earth's natural resources is one of the most pressing environmental problems facing the world we inhabit. The promise and challenge of **resource management** is well-illustrated in terms of water: its **availability, use**, and the **competing demands** for this natural resource in states like Oregon.

Oregon is blessed with some of the finest natural sites in the nation: a Pacific coastline of unparalleled beauty; the spectacular **forest** wilderness of the Cascade Range; the Columbia River and the gorge it creates as it transects the Cascades; and the **high desert** environments found in the eastern part of the state. Its citizens are well known for their environmental consciousness, and the state's legacy of "national firsts" is impressive: comprehensive statewide land-use planning; air quality regulations; a recycling law to reduce solid waste in landfills; public ownership of coastal areas; and private-sector volunteer campaigns to maintain the natural beauty of coastal areas.

The backdrop for the state's natural beauty can be discerned through a description of the north Pacific Ocean's effect on the state's topography as described in the **rain shadow** explanation in the Discussion of Case Study Themes.

The **population distribution** of Oregon follows the distribution of water resources. Most people are concentrated on the western side of the Cascades, from Portland south through the Willamette Valley, and secondarily along the east-west trending portions of the Columbia River. Demand for water resources, however, while concentrated in the population centers in the western one-third of the state, is also high in the sparsely settled but agriculturally important section of northeastern Oregon, where wheat production is concentrated.

Understanding the geography of water and its wise utilization is a critical challenge facing Oregon. Even in a state of renowned environmental consciousness, balancing the options is a complex exercise. While the solutions to water resource management problems are far from simple, the geography lessons are clear. The **location of resources** like water influences the distribution of people and their activities. Technology, in the form of multipurpose dams for **hydroelectric power generation** and **irrigation,** changes how humans appraise natural resources, and it may modify the ways in which resources have traditionally been used. Technology may also generate new economic systems associated with resource use. **Demand** for water resources often varies spatially, across the face of the globe as well as within individual countries and states. **Conflict** over resources is likely to increase as demand increases, either due to population growth or land-use activities, such as agriculture, that are **water intensive.**

To further develop your understanding of orographic precipitation, refer to the box The Rain Shadow Effect in Chapter 3, "North America: The Postindustrial Transformation," from de Blij and Muller, 1994, *Geography: Realms, Regions, and Concepts: 7th Edition.*

CRITICAL VIEWING

Based on the Case Studies in Video Program 11: "Regions and Economies" from *The Power of Place: World Regional Geography,* and the reading in Chapter 3, "North America: The Postindustrial Transformation," from de Blij and Muller, 1994, *Geography: Realms, Regions, and Concepts: 7th Edition,* develop essays answering the following questions:

1. Discuss water resource management conflicts between farmers and Native Americans in Oregon. How might you advise the Governor of Oregon on ways to balance the competing interests of each group for this scarce resource?

2. The salmon runs of the Pacific Northwest are dwindling at an alarming rate. Are there ways to balance the economic needs of urban and agricultural interests, the ecological needs of the salmon, and the rights of Native Americans?

3. Discuss the reasons Native Americans are hesitant to sue farmers in eastern Oregon who may be using water for irrigation without legal authorization?

4. International trade is an increasingly vital dimension of the U.S. economy. Agricultural exports are one example of Oregon's comparative advantage in international trade. How might Oregon's position on free trade and open international markets differ from those Midwestern states that now face increased competition from Japanese automobile manufacturers?

5. Diffusion of ideas and innovations is a theme common to both *U.S. Midwest: Spatial Innovation* in this Video Program and *Poland: Diffusion of Democracy* in Video Program 4. Compare and contrast the explanation of innovation theory presented in these two Cases Studies. What are the "barriers" and the "carriers" either facilitating or impeding the diffusion of ideas and innovations found in each analysis?

■ UNIT 5

The Geographic Dynamics of the Western Pacific Rim

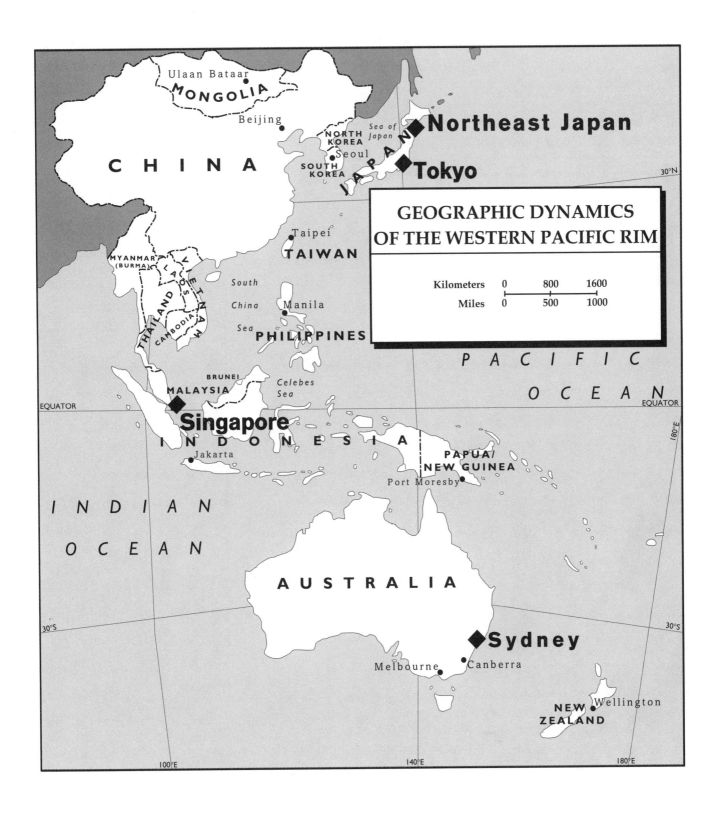

GEOGRAPHIC DYNAMICS
OF THE WESTERN PACIFIC RIM

Kilometers	0	800	1600
Miles	0	500	1000

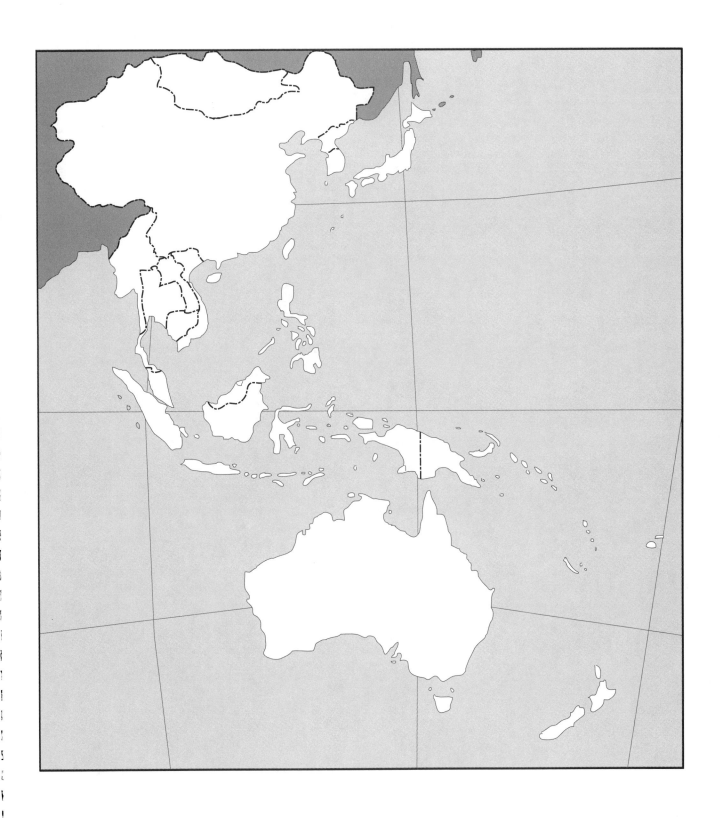

■ PROGRAM 12

The Japanese Paradox: Small Farms and Mega-Cities

LESSONS IN GEOGRAPHY

At the end of Program 12, you should be able to:

1. Explain the influence of climate on agricultural activities in Northeast Japan.
2. Identify strategies to respond to constraints placed on rice farmers by the physical environment.
3. Review the geographic results of policies and programs for resource use and management vis-à-vis labor in Japanese agriculture.
4. Discuss how the role of transportation technology shapes the characteristics and growth of Tokyo.
5. Understand Tokyo's meaning and significance for Japan's place in the world economy and the process of globalization.
6. Understand how the dynamic relationship between agricultural and urban Japan leads to the formation of places and to a sense of personal and community identity.

OVERVIEW

Program 12 highlights a modern paradox in Japan — the coexistence of both the mega-city of Tokyo and small-scale agriculture. In *Northern Japan: Protecting the Harvest* the farmer's story illustrates climatic conditions in *Tohoku,* or Northeast Japan, and the influence of **natural hazards** upon **agricultural productivity.** In *Tokyo: Anatomy of a Mega-City* a commuter from Saitama provides insight into the continued growth of one of the world's largest metropolitan areas.

Rice, the staple of the Japanese diet, has played a critical role throughout history in the culture as well as in the contemporary politics and economy of Japan. This Case Study illustrates wet-rice production on the northeastern edge of the main island of Honshu. In this **marginal area,** winter finally ends in May and signals the time to transplant rice plants that have been forced to sprout in hot houses. The seedlings grow in fields of standing water, exposed to seasonal weather changes. To overcome the **adverse climatic factors** affecting rice production requires both advanced technology for creating agricultural machinery and predicting weather patterns and intensive labor. A farmer from the village of Rokunohe manages his crop based upon seasonal conditions, irrigation and drainage needs, transplantation, and harvest.

The farmer's worst fear in Northeast Japan is *yamase,* a dense fog accompanied by cold, east winds, which can cause heavy damage to rice crops. This Case Study presents the **atmospheric conditions** responsible for creating this adverse weather and illustrates how modern research is aiding agriculture in this area of Japan. The final section of the Case Study provides insight into Japanese agriculture as the labor force declines in this sector, moving to take advantage of higher-paying opportunities now available in rural areas.

The Tokyo study reviews how the **transportation system** has contributed to **urban growth** and the implications for Tokyo's future. With a metropolitan population of over 30 million, the role of greater Tokyo's **infrastructure** is viewed through the eyes of a commuter, illustrating that the transportation system is one component that helps the metropolis to function well despite its size.

To view the city's immense transportation infrastructure, the program examines Otemachi, the largest subway station in Tokyo. The station serves commuters who travel from outlying areas to work in Tokyo due to high land costs and a lack of affordable housing in the city. With over 1 million people working in the area of the station, most of them commuters, the Otemachi area has two populations — day and night.

Tokyo's role in Japan is unique, serving as Japan's capital and largest city. With one out of four Japanese residing here, most of Japan's economic, political, and legal activities are concentrated within its metropolitan borders. Despite the city's large population, it continues to grow and prosper. This expansion can only take place supported by a comprehensive transportation network. Covering a land area only slightly larger than Chicago, the city functions with three times Chicago's **population density**. This program portrays Tokyo's role as one of the great cities of the world, a key player in the world economy and a forerunner in the process of **globalization**.

STUDY RESOURCES

1. Video Program 12: "The Japanese Paradox: Small Farms and Mega-Cities" from *The Power of Place: World Regional Geography.*

2. Unit 5 Map, "Geographic Dynamics of the Western Pacific Rim" from Latz and Gilbert, 1996, *The Power of Place: World Regional Geography Study Guide.*

3. Chapter 4, "The Pacific Rim of Austrasia" from de Blij and Muller, 1994, *Geography: Realms, Regions, and Concepts: 7th Edition.*

PREVIEWING QUESTIONS

1. Why does Japan devote so much energy to scientific and technological research in growing rice in northeastern Japan?

2. What physical/climatic geographic features influence Northeast Japan?

3. How can the Tokyo metropolitan area, home to approximately 32 million people, continue to grow?

4. How does Tokyo's transportation system differ from that found in a North American city, for example, New York, Chicago, or Los Angeles?

KEY THEMES

- **Natural Hazards and Agricultural Productivity**
- **Rice Agriculture in a Northern Climate**
- **Growth of Tokyo**
- **Tokyo's Transportation System**
- **Future Trends**

DISCUSSION OF CASE STUDY THEMES

• Natural Hazards and Agricultural Productivity

Environmental factors pose a great risk to farmers both in Japan and throughout the world. Climatic conditions are factors beyond human control which affect agricultural activities and production. As mentioned earlier, the rice farmer of northeastern Japan faces the risk of unseasonably cool weather which can affect the rice crop at the critical early phase of plant development.

Techniques such as water control and fertilizer management are important steps which can be taken to modify the adverse conditions created by yamase. Advanced weather forecasting assists the farmer who must then respond quickly to rapidly changing weather conditions. The process is extremely labor-intensive and requires constant vigilance. As forecasting improves, the vigilant farmer has more time to ensure a successful harvest.

Modernization in Japan has had an effect on agriculture throughout the country. Fewer farmers can afford to work full time in a rural environment demanding intensive labor. Despite high government subsidies to farmers, including a guaranteed market price for rice as well as infrastructure investment projects, the higher pay in cities and better non-agricultural job opportunities in rural areas have increasingly attracted farmers away from a full-time presence in the fields. Part-time farmers who split their time between the office and the field are not able to devote the significant time required to intensely manage the rice crop along the northern edge of its range.

• Rice Agriculture in a Northern Climate

Rice, the staple food of Japan, is a marshland plant of tropical origin. For over 2,000 years, the cultivation of rice has been traditionally regarded as a religious act and is symbolically linked to Japanese culture.

The northern latitudes of Honshu and Hokkaido are the farthest northern exposures that rice crops can endure. Several thousand varieties of rice are grown in Japan. Hybridization, fertilization, and increased productivity due to advances in irrigation and drainage have significantly raised yields in all areas.

Rice grows best in warm temperatures. Rice will not germinate below 46-50°F (8-10°C), and ideal conditions for growth are high temperatures, 79-88°F (26-31°C). The plants are resistant to severe oxygen shortage but cannot tolerate drought.

The wide range of climates affecting rice agriculture is due to Japan's relative location. Growing conditions are influenced by the Eurasian continent to the west and Japan's linear orientation from north to south, creating both subarctic and subtropical climatic zones. The second major influence is Japan's location in relation to the temperate monsoon zone of East Asia. The winds of the monsoons largely determine Japan's seasons. The physical geography of Japan is characterized by high mountains, deep valleys, narrow plains, and moderation by Kuroshio or the Japan Current, the warm Pacific ocean current to the east. Combined, these conditions create wide seasonal variation and dramatic weather changes.

Yamase occurs when the normally strong North Pacific high is weakened by a stronger Sea of Okhotsk high pressure zone centered east of the Kurile Islands. Wind patterns between pressure zones flow from strong to weak systems and, in this Case Study, from north to south. This phenomenon causes the cold air mass to move south, dropping temperature below 68°F for days or even a month at a time in the summer.

• Growth of Tokyo

Tokyo is already one of the most densely settled areas of human activity in the world, and it continues to grow. The concentration of the country's financial, government, information, and industrial sectors forms the leading economic center in the country, making Tokyo a major player in the globalization of the world's economy. This dynamic economy competes successfully in world markets and attracts residents seeking work opportunities.

Within Japan, Tokyo is distinguished from other cities due to its location atop the largest alluvial plain the country. This relatively large level area, the Kanto Plain, allows for an overwhelming concentration of government agencies, cultural organizations, universities, research organizations, newspapers, television stations, publishers, and communication businesses.

Tokyo's locational attributes, desirable for businesses on a regional and international scale, create a high demand for land which is reflected in extremely high prices for real estate, making it largely unaffordable for residential use. Most of the labor force required for a service and industrial economy must reside in the city's periphery due to limited housing in the city center.

• Tokyo's Transportation System

The ability of the city not only to function but to grow rests with an unmatched transportation system. The system is designed to converge on Tokyo's city center, and the rail lines radiate from the central core. People in Tokyo have been forced to dwell farther and farther from the city center due to high land costs. Surveys show that only 40,000 people actually reside within the core of Tokyo, depicted in this Case Study as the central business district; however, the work-related population increases by 1 million people daily. These commuters must travel for as long as two hours each way every day. The transportation network for moving such large numbers of people is particularly dense in a zone stretching 19-22 miles (30-35 kilometers) to the north, south, and west of Tokyo. Commuters trade long commutes for affordable housing and amenities such as clean air and yard space found in the metropolitan periphery. As seen in the video, different routes may increase rider comfort but may also lengthen travel times.

The efficiency and reliability of the train system has provided workers with access to both affordable housing in outlying areas and the urban job market. The transportation system, land costs, and workers' preferences have influenced the horizontal and transport arterial-based growth pattern of Tokyo and other urban centers in Japan.

• Future Trends

The transport service provided by the rail system can serve only a limited number of people. Increased demand has led to the construction of new rail lines, but many corporations are also in the process of developing plans to relocate from the urban core to the suburbs, reducing commuting time by relocating where workers reside. The tremendous concentration of buildings, transportation infrastructure, and businesses in Tokyo acts as a drag on the outward-directed relocation process; the effort to reduce congestion has been very slow.

New urban plans are attempting to address the problems of commuting. Alternative housing projects, such as Makuhari New Town depicted in the film, are underway near new commercial sites along the waterfront of Tokyo Bay. The high demand for such housing reflects consumer preference, a factor which once again may change the growth pattern of the future.

Expansion of intra-regional trade and Tokyo's role in the world economy also influences the city's future. The Asia-Pacific region is now poised to become the world's dominant regional economy due not only to its size but also to its expected growth rate. Within this regional economy, Japan contributes 75 percent of the total production, mirroring the relationship of the U.S. economy to the Western Hemisphere. In the past few years, devaluation of the U.S. dollar and Japanese recession have tempered the Japanese economy in the Asia-Pacific region and in the world without fundamentally altering the general regional economic trends. Increasing global demand for Japan's products and services strengthens the region's economy and Japan's relative locational advantage.

CASE STUDY CONNECTIONS

1. Northeast Japan is located in a colder and more unpredictable environment than Tokyo, which is located in a more temperate climate.

2. There is a vast difference of scale between the agricultural farming areas and the huge cities of Japan.

3. The work available in the urban environment and the secondary and tertiary sectors attracts labor from rural locations. This profoundly affects regions like the northeast which are on the margins of the rice-growing range and require intensive, knowledgeable labor.

4. The agricultural process of rice production represents a traditional practice that is being marginalized as modernization continues.

5. Japan has expended tremendous resources and even expanded into marginal areas in order to become self sufficient in rice production; even so, demand for rice has decreased in urban areas.

6. Population density is a major difference between the areas.

7. The inhabitants of both regions provide a contrast in lifestyle that captures the urban/rural paradox of modern Japan.

8. Governmental power is concentrated in Tokyo; and policy development there greatly influences the structure of agriculture dominant in Northeast Japan.

9. Globalization has affected both urban and rural areas, but Tokyo remains at the forefront economically in the country and in the world.

GEOGRAPHERS AT WORK

Kanno Hiromitsu is an atmospheric geographer. He studies weather patterns in Northeast Japan.

To better understand yamase, Kanno explains that warm summer conditions are associated with the dominance of a Northern Pacific high pressure system, and winter conditions are associated with dynamic interaction between a Siberian high pressure system, centered just north of Korea on the Eurasian continent, and the Aleutian low in the Pacific Ocean east of Honshu.

Kanno researches better ways to predict yamase so that farmers who can no longer devote day-and-night attention to their fields have the best advance warning possible to protect their crops.

TEST YOUR UNDERSTANDING

Questions

1. Describe the cultural preferences and market forces which have influenced the growth pattern of Tokyo.

2. Within Japan, what distinguishes Tokyo from other Japanese cities?

3. What other factors besides climate affect rice production in Northeast Japan?

Map Exercises

The maps provided at the beginning of this unit can be used to complete this exercise. You may find extra copies of these maps helpful. Also, refer to the maps of Chapter 4, "The Pacific Rim of Austrasia," from de Blij and Muller, 1994, *Geography: Realms, Regions, and Concepts: 7th Edition*, and any standard atlas to assist in this exercise.

1. Place the following names:

Japan Current (Kuroshio)	Sea of Japan	Kanto Plain
Sea of Okhotsk	Tokyo Bay	Tokyo
Tsugaru Strait	Tohoku	Hokkaido
Rokunohe	Kawasaki	Honshu
Aomori Prefecture	Saitama Prefecture	Kurile Islands

2. Map the major train routes which originate from Tokyo.

   How is Aomori Prefecture linked to Tokyo?

   Why is the system so highly developed around Tokyo?

3. Label the following atmospheric conditions:

Normal winter positions of the Siberian High and the Aleutian Low.

Mid-summer position of the North Pacific and Okhotsk high systems.

CONSIDER THE DISCIPLINE

The Geography of Development

Japan is a country that has become modernized within the last 125 years, relatively recently compared to the industrialized countries of the West. In the course of Japan's **industrialization,** it has sought to modernize certain traditional features of economy and society in both the agricultural sectors and the large urban areas which have characterized the country since the late 19th century.

Japan's geographical situation gave it a modest but adequate base of resources with which to modernize, e.g., coal reserves for industry and a well-developed agricultural infrastructure for growing rice. Yet, the country's **resource endowment** could not sustain the prodigious economic output of the present day — Japan is second only the United States in economic output per capita — without careful management of its resources and efficient use of **mass transportation.** Considering that the **population** of Japan increased threefold between 1868 and 1968, the country's non-Western modernization is a compelling study of **20th century development**. Such successful assessment and management of resources by the world's seventh most populous country is distinctly Japanese: an intimate knowledge of the environment of the island country as well as the cultural attitude and willingness to make sacrifices necessary to achieve lofty economic and social goals.

To further develop your understanding of this field of regional geography, refer to the Focus on a Systematic Field in Chapter 4, "The Pacific Rim of Austrasia," from de Blij and Muller, 1994, *Geography: Realms, Regions, and Concepts: 7th Edition.*

CRITICAL VIEWING

Based on the Case Studies in Video Program 12: "The Japanese Paradox: Small Farms and Mega-Cities" from *The Power of Place: World Regional Geography,* and the reading in Chapter 4, "The Pacific Rim of Austrasia," from de Blij and Muller, 1994, *Geography: Realms, Regions, and Concepts: 7th Edition,* develop essays answering the following questions:

1. What are some of the ways modern and traditional societies coexist in Japan?

2. Why did Japan's modern industrialization depend on the acquisition of raw materials from distant locales?

3. How has Japan's areal functional organization developed based on regional specialization?

4. What characteristics of Japanese modernization are non-Western?

▪ PROGRAM 13

Global Interaction

LESSONS IN GEOGRAPHY

At the end of Program 13, you should be able to:

1. Identify the locational advantages of Singapore which have fostered economic development.

2. Examine the role of Australia and its contributions to trade and development in Pacific Asia.

3. Review the historic and cultural factors which have influenced the spatial distribution of Australia's population.

4. Identify the factors allowing Singapore to become a global distribution hub.

5. Understand the role of immigration in modern Australian and Singaporean society, and its past pattern and historic influence.

OVERVIEW

Program 13 examines two countries, Australia and Singapore, located in the Asia-Pacific region. Both nations have achieved advanced development and offer high living standards for their respective populations. Together they are a study in geographical, cultural, and historical contrasts.

highest incomes

In *Singapore: Gateway City* locational advantage is discussed from historical and modern perspectives. Singapore now serves as a global **distribution hub** in the world's economy, and this Case Study provides insight into the factors that have turned this city-state into one of the world's busiest ports. Factors contributing to the rise of this **newly industrializing economy** include Singapore's **strategic location**, modern cargo handling facilities, and skilled work force. Computer technology, advanced information management, and telecommunications are components discussed in the video which comprise a **modern port facility**, important to Singapore's success as a **trade entrepôt.**

In *Australia: New Links to Asia* the historic role of industry in the development of the economy is examined as a factor in creating the country's highly **urbanized population**. Marked by a shift in economic trade from Europe to Asia, the effects of this dramatic change on modern Australian society are examined. Australia, once settled by Europeans, is now a destination for Asian **immigrants** as well.

Economic trade in Australia remains based upon the export of **natural resources,** while international finance, telecommunications, and information services are rapidly becoming established. Originally favoring the export of wool and agricultural commodities, Australia is also noted for its rich mineral and energy resources. The country possesses minerals such as iron ore, nickel, bauxite, precious metals, gemstones, and non-metallic minerals used in industry.

STUDY RESOURCES

1. Video Program 13: "Global Interaction" from The Power of Place: World Regional Geography.

2. Unit 5 Map, "Geographic Dynamics of the Western Pacific Rim" from Latz and Gilbert, 1996, The Power of Place: World Regional Geography Study Guide.

3. Chapter 4, "The Pacific Rim of Austrasia," from de Blij and Muller, 1994, Geography: Realms, Regions, and Concepts: 7th Edition.

PREVIEWING QUESTIONS

1. Where are Singapore and the major cities of Australia located and what common characteristics do they share?

2. What is the largest city in Australia?

3. How do the countries of Australia and Singapore differ?

4. What are the economic bases of Singapore and Australia and how are their economies related?

5. How has immigration affected the composition of each country's population?

KEY THEMES

- **Historical Influences**
- **Singapore's Locational Advantages**
- **The City-State's Role as a Global Distribution Hub**
- **Components of a Modern Port Facility**
- **Australia's Population Distribution**
- **The Shift in Economic Orientation from Europe to Asia**
- **Immigration**

DISCUSSION OF CASE STUDY THEMES

• Historical Influences

Serving as a cultural crossroads since the 14th century due to its excellent harbor, Singapore's population reflects a diverse cultural geography of Chinese, Indian, and Malay peoples. Over time, trade and the introduction of ideas from different cultures have resulted in a modern city-state of great contrast and variety. British development of the port of Hong Kong reduced trade in the mid 1800s, but the opening of the Suez Canal in 1869 coupled with the advent of steamships began an era of prosperity. Western demand for rubber and tin produced on the adjoining Malay peninsula were important in Singapore's development into a major port. A legacy of British colonialism can be found in the country's governmental structure, and the early development of its port and many historic buildings provides evidence of past human migrations. British colonialism ended in 1963 when Singapore briefly became part of the Malaysian Federation.

(handwritten margin notes: incorruptability + politically stable)

• Singapore's Locational Advantages

Even before Singapore's secession from Malaysia in 1965, this entrepôt had become the fourth busiest in the world, based on the number of ships served. Situated between the Strait of Malacca and the Strait of Singapore, its deep natural harbor provides a critical port linking the Indian Ocean with the South China Sea. This strategic location has served a role in commodity transactions for centuries among Southwest, Southeast and East Asia, especially Indonesia, China and Japan. In today's global economy, Singapore is centrally located with respect to Asia and the rapidly growing Pacific Rim, thus providing efficient and cost-effective connections for the international business community.

• The City-State's Role as a Global Distribution Hub

(handwritten margin notes: location, govt, infrastructure, workforce (skilled), lid, to distrib distr)

A combination of factors including relative location, government policy, modern infrastructure, and a highly skilled work force have contributed to Singapore's role as the key regional distribution center linking the world to the Asia-Pacific region. The tightly controlled government, dominated by a single political party since 1959, has fostered an extremely efficient port facility through development policy. This policy and capital investment funded by the government have allowed Singapore to develop advanced sea and air facilities which compete worldwide. The modern port and the establishment of a free trade zone, avoiding import and export duties, have attracted international businesses like Hewlett Packard who center regional operations in Singapore, distributing their products to smaller ports throughout the region. Singapore's economic success rests with its modern seaport and airport facilities, stable financial structure and banking institutions, information technology, and its skilled pool of human capital.

• Components of a Modern Port Facility

(handwritten margin notes: modern + efficient)

To compete worldwide as a transportation center and to attract new businesses, extremely efficient technological and human management systems are required. A modern infrastructure consisting of equipment, storage facilities and information technology is also necessary. Efficient computer-assisted container facilities at Singapore's seaport, in tandem with the air freight center, distribute products regionally as well as to Europe, the United States, and Australia. Service industries such as information management, finance, and telecommunications have developed to sustain the port facilities and the businesses operating in them.

• Australia's Population Distribution

Eighty-five percent of Australia's population lives in towns and cities, making it one of the world's most urbanized societies. This concentration of the population in urban areas is a reflection of the development of Australia's economy. Driven by the export of commodities and capital investment from Britain, Australian cities developed into centers of commercial activities, and job opportunities became highly concentrated in urban areas. In contrast, far fewer industries developed throughout Australia's vast interior. The rail system necessary to move natural resources and commodities allowed central business districts (CBDs) to develop near ports where goods were shipped overseas. The convergence of railroads from outlying areas played a key role in the growth of the city. Now the automobile has replaced the train as the public's main mode of transportation and has helped to create the sprawling suburbs of single family homes favored by modern Australian society.

In the video, Sydney is examined as an example of a major urban area in Australia. Chosen in 1788 as the location of the first British colony on Australian soil, Sydney reflects the development of this settlement. Prospering early due to a strong wool trade and excellent natural harbor, successful businesses developed in the 1830s, serving the newly industrialized Britain in the 19th century. The introduction of technology, along with the infusion of British capital, created commercial centers, the foundation of the city of Sydney and other capital cities of the British colonies in Australia.

• The Shift in Economic Orientation from Europe to Asia

Australia's economy was primarily based upon agriculture until the end of World War II. Beginning with the wool industry, as reviewed in the video, it later expanded into wheat, beef, dairy, and irrigated crops. Economic diversification began at the end of the war with the growth of the manufacturing and service industries. The exploitation of minerals significantly accelerated economic growth. This pattern of development remained dependent upon foreign investment and capital and was highly vulnerable to fluctuations of commodity prices. The manufacturing sector has remained dependent upon foreign interests and has suffered a slow decline which began in the 1960s.

In the 1990s, Australia's economy continues to be based upon the export of commodities, and the role of shipping has remained very important. Commodity exchange is now directed toward Asia, a significant change from Australia's historically European-based trading partners. Evidence of this shift can be found in analyzing financial transactions, capital investments, air transportation routes, and expanding telecommunications systems. All indications point toward Asia, notably Japan and the newly industrializing economies of the Asia-Pacific region.

• Immigration

Australian society has traditionally been regarded as British or Anglo. This has been generally true with the exception of small groups of Chinese, Germans, and other largely European ethnic groups who formed small concentrations beginning in the 19th century. Ethnic diversification has remained a controversial issue in Australia, and policy favoring British and Irish immigration persisted for decades in the 20th century. Major world crises of the 20th century, particularly from the 1950s through the 1980s, have produced fresh waves of immigrants from Greece and Italy, and most recently from Hungary, the former Czechoslovakia, Lebanon, Turkey, Chile, Laos, Vietnam, Cambodia, and China. When government policy favoring white immigrants was formally abolished in 1973, increasing numbers of immigrants began arriving from Asia. By the early 1980s, people of overseas origin accounted for almost 30 percent of the populations of Sydney and Melbourne, and 22 percent of the country.

The shifting and dynamic economy of the Asia-Pacific region in the 1990s has helped to create opportunities for employment in Australia, a key component of human migration. Modern Australia is experiencing pronounced changes in its population based upon the immigration of Asian immigrants as reviewed in the example of Fairfield, a suburb of Sydney. Asian immigrants are a significant component of the Australian economy, marking a shift away from Europe. Asians not only represent sources of labor, skills, and capital; Asian contacts provide access to overseas markets important to business. These demographic changes reflect the economic transformation underway in modern Australia.

CASE STUDY CONNECTIONS

1. Both nations are dominated by the influence of important urban areas with excellent natural harbors.

2. Both nations were past colonial possessions of the British Empire.

3. Singapore's economy was never dependent upon the production and export of commodities, while Australia's was built upon that very system.

4. Australia possesses a vast natural resource and land base, while Singapore has a limited resource base and occupies a mere 250 square miles (647.5 square kilometers).

5. Both countries have important histories of immigration, with Singapore's dating back to the 14th century, and Australia's primarily occurring from the 18th through the mid-20th century.

6. Australia's population remains relatively homogenous but has begun to change with the arrival of Asian immigrants; Singapore's population composition reflects considerable diversity but is dominated by the Chinese.

7. Singapore lies within a single, uniform climatic zone, while Australia covers a wide range of climates due to its continental proportion in latitude and longitude.

TEST YOUR UNDERSTANDING

Questions

1. Why is Australia's population so urbanized despite its vast size? List three reasons.

2. List four characteristics that make Singapore a global distribution hub.

3. Explain how Singapore has developed into the region's global distribution hub.

4. Why is air transport such an important indicator in assessing Australia's economic activity?

Map Exercises

The maps provided at the beginning of this unit can be used to complete this exercise. You may find extra photocopies of these maps helpful. Also, refer to the maps of Chapter 4, "The Pacific Rim of Austrasia," from de Blij and Muller, 1994, *Geography: Realms, Regions, and Concepts: 7th Edition*, and any standard atlas to assist in this exercise.

1. Place the following names:

Sydney	Hong Kong	Equator
Singapore	Indian Ocean	Tropic of Cancer
Melbourne	Strait of Singapore	Tropic of Capricorn
Indonesia	International Date Line	Perth
Jakarta	South China Sea	Malay Peninsula
Bangkok	Strait of Malacca	Java Sea

2. Map at least five agricultural and eight mineral resource locations on the Australian continent. Outline four possible shipping routes originating from Singapore's harbor

CONSIDER THE DISCIPLINE

The Geography of Development

ECONOMIC INTERACTION AND DEVELOPMENT

The development of Singapore and Australia is linked to each country's historic ties with different cultures and the **relative location** of their harbors, natural resources and sources of capital for investment. Each portrays different **patterns of development.** Singapore provides an example of a classic entrepôt, while Australia remains dependent upon the export of commodities and raw materials. Examples of different development factors include a lack of natural resources and land area for Singapore, and remoteness and arid environment for Australia. Each has overcome significant disadvantages to develop an advanced economy and high living standard for its respective population.

As the world shifts to a more competitive **global economy**, trade is driven by efficiency as well as cultural ties. Intra-regional trade within the western Pacific Rim is enhanced by efficient air and sea transport, excellent telecommunications, and the absence of significant time zone differences despite vast distances between markets. The transition from commodity export to service and high-technology industries remains a significant barrier facing Australia. Singapore, despite problems created by a more affluent middle class demanding greater individual autonomy, remains poised for continued growth. This position is a result of its modern **port facilities** which enable adjustment to shifting market forces associated with the change from traditional movement of commodities to highly valued technological products.

THE EFFECTS OF IMMIGRATION

Both countries have deep historic ties to cultures located overseas; Australia's roots lie in Europe, primarily in Britain, while Singapore is linked to Asia. The distinction between the two is rooted in history: Singapore's **immigration patterns** date back to the 14th century; Australia has had more recent European immigration. Economic opportunities have fueled human migration over time creating a rich and diverse mixture of cultures reflecting both countries' modern societies.

Modern human migration continues today with patterns of immigration associated with the dynamic Asia-Pacific economy. The majority of immigrants relocating to Australia in the late 1980s and today arrive from Asia. These immigrants originate in Hong Kong, Vietnam, China, the Philippines, Malaysia, Laos, India, and Sri Lanka. Singapore's immigrant population includes people from neighboring Malaysia and Indonesia; the country has recently opened its doors to Hong Kong hoping to entice Chinese capitalists. The cultural mixture will profoundly influence each country's future society, resulting in further **diversity** and new ties to overseas enclaves. Economic development is likely to mirror these patterns of **human networks**.

To further develop your understanding of this field of regional geography, refer to the Focus on a Systematic Field in Chapter 4, "The Pacific Rim of Austrasia," from de Blij and Muller, 1994, *Geography: Realms, Regions, and Concepts: 7th Edition.*

CRITICAL VIEWING

Based on the Case Studies in Video Program 13: "Global Interaction" from *The Power of Place: World Regional Geography,* and the reading in Chapter 4, "The Pacific Rim of Austrasia," from de Blij and Muller, 1994, *Geography: Realms, Regions, and Concepts: 7th Edition*, develop essays answering the following questions:

1. Which historic factors have influenced Singapore's development into the world's fourth busiest port (by number of ships served)? Has the diverse culture of Singapore's population contributed to its success and if so, how?

2. Explain why Australia has lagged behind economically when compared to the developed countries of the Asia-Pacific region.

3. Define the Pacific Rim. What roles have immigration and culture played in the development of this concept?

■ UNIT 6

Middle America: Collision of Cultures

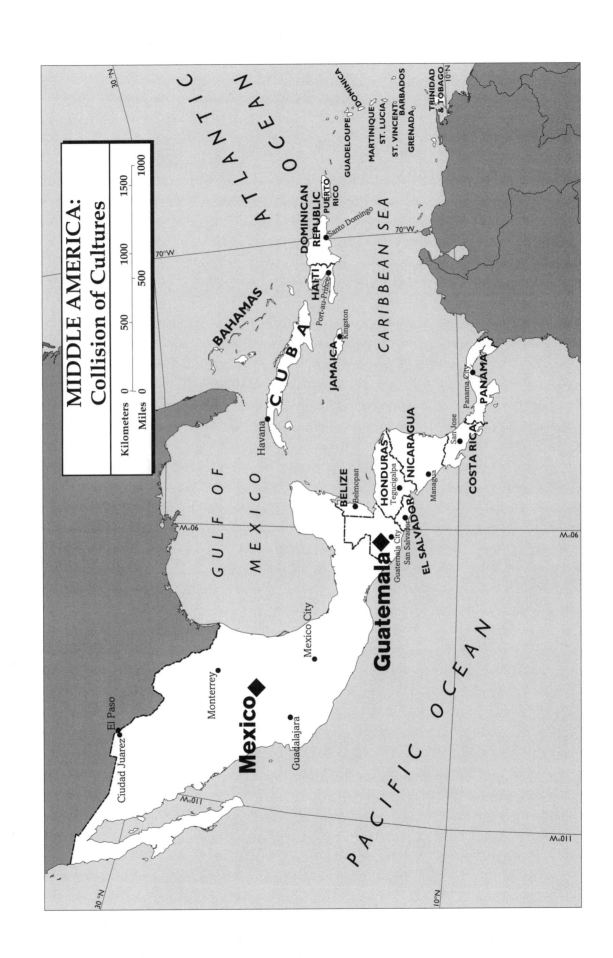

MIDDLE AMERICA:
Collision of Cultures

Kilometers 0   500   1000   1500   1000

Miles 0   500   1000

ATLANTIC OCEAN

BAHAMAS

CUBA

Havana

GULF OF MEXICO

Mexico ◆

Ciudad Juarez

El Paso

Monterrey

Guadalajara

Mexico City

JAMAICA

Kingston

HAITI

Port-au-Prince

DOMINICAN REPUBLIC

Santo Domingo

PUERTO RICO

CARIBBEAN SEA

GUADELOUPE

DOMINICA

MARTINIQUE

ST. LUCIA

ST. VINCENT

BARBADOS

GRENADA

TRINIDAD & TOBAGO

BELIZE

Belmopan

Guatemala ◆

Guatemala City

HONDURAS

Tegucigalpa

EL SALVADOR

San Salvador

NICARAGUA

Managua

COSTA RICA

San Jose

PANAMA

Panama City

PACIFIC OCEAN

30°N

70°W

10°N

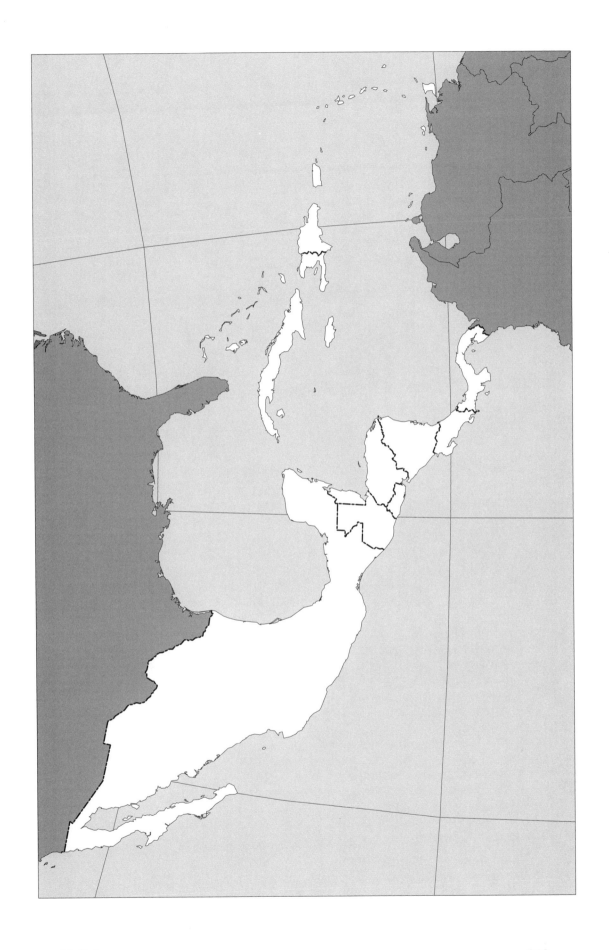

▪ PROGRAM 14

Migration and Conquest

LESSONS IN GEOGRAPHY

At the end of Program 14, you should be able to:

1. Identify spatial patterns which characterize the points of origin of illegal Mexican immigrants to the United States.

2. Examine the causes of migration from Mexico's "Hollow Core," the Mesa del Norte.

3. Explain how the growth of Mexico's maquiladoras is the result of a government policy to increase employment in manufacturing throughout the country.

4. Understand the growth of Monterrey, Mexico's third largest city and the capital of the Nuevo León state, home to many of Mexico's maquiladoras.

5. Describe the physical geography of Guatemala in relation to its cultural and economic landscape.

6. Discuss the role that land ownership and land tenure issues play in the political and social systems of Middle America.

OVERVIEW

Unit 6 explores some of the key themes which affect the realm of Middle America.

The first Case Study, *Mexico: Motive to Migrate,* explores **migration** both within Mexico and to Mexico's northern neighbor, the United States. By examining **immigration** records, a pattern of departure from Mexico's Mesa del Norte can be identified. This arid plateau has a poor, agricultural economic base and a depressed silver mining economy. Migration to the United States is common among the people of the rural town of Cedral, located in the heart of the Mesa del Norte, though many migrants return to their homelands after a season or a year in the U.S. But not all migrants in Mexico are headed to the United States. The city of **Monterrey**, the capital of the border state of Nuevo León, has recently experienced a large population influx and, in the past 15 years, has grown from 1.7 million to 2.8 million people. One of the reasons for this growth is the labor demand created by the expansion of Mexico's manufacturing industries, or **maquiladoras**.

Maquiladoras are the result of a government program to expand Mexico's role in **international trade**. Industries relocate to Mexico in exchange for tariffs on the value-added portion of products shipped out of the country. By shipping the parts of a product to maquiladoras to complete assembly, foreign companies are helping to invigorate the economy of the border region. As the **North American Free Trade Agreement (NAFTA)** takes effect, migration patterns in Mexico will once again be altered. One result of NAFTA will be the end of governmental **price supports** for Mexico's **agricultural sector**. But, even as this further impoverishes farmers in the

Mesa del Norte, new opportunities will arise as the maquiladoras of northern Mexico decentralize into the north central region, including the community of Cedral. This will help to diversify the economy of that region, decrease the unemployment rate, and decrease migration out of the community, thus keeping families together.

The second Case Study, *Guatemala: Continuing Conquest,* examines the **historical geography** of Guatemala. The country's history of Spanish **conquest** is examined beginning in 1524, when Mayan mountain Indians were subjugated, not for their land or wealth, but to provide **labor** needed in order to create and maintain **plantations** and **haciendas**. After Spanish rule was overthrown in 1821, the country lapsed into a period of political turmoil until 1873 when Justo Rufino Barrios **redistributed the lands** of the Roman Catholic Church and revived the plantation economy. The Mayan Indians were once again exploited as a source of labor, this time by international businesses intent on producing crops such as coffee, sugar, and bananas. The political status of the Mayans once again deteriorated, while **land ownership** became more concentrated, with 3 percent of the population controlling two-thirds of the land. A second attempt at land reform in the 1950s led to a U.S.-backed invasion that overthrew the liberal-democratic government of Jacobo Arbenz Guzman, touching off a civil war. The country continues to experience civil unrest, and historic homelands of the Mayan Indians are now occupied by the Guatemalan military. As the revolt continues, the military tries to crush the opposition and its suspected sympathizers by means of state terrorism.

STUDY RESOURCES

1. Video Program 14: "Migration and Conquest" from *The Power of Place: World Regional Geography*.

2. Unit 6 Map, "Middle America: Collision of Cultures" from Latz and Gilbert, 1996, *The Power of Place: World Regional Geography Study Guide*.

3. Chapter 5, "Middle America: Collision of Cultures" from de Blij and Muller, 1994, *Geography: Realms, Regions, and Concepts: 7th Edition*.

PREVIEWING QUESTIONS

1. Where were the conquistadores from and when did they land in Middle America?

2. What is the buying power of a Guatemalan quetzale? A Mexican peso?

3. How are haciendas and plantations similar? How do they differ?

4. What is a maquiladora?

KEY THEMES
- **Push Factors of the Mesa del Norte**
- **Migration to Monterrey's Maquiladoras**
- **The Future of the Mesa del Norte**
- **Spanish Conquest in Guatemala**
- **Plantation System Imposed**
- **Land Alienation Leads to Violence**
- **Military Occupation**

DISCUSSION OF CASE STUDY THEMES

• Push Factors of the Mesa del Norte

Most migrants in and from Mexico are from the Mesa del Norte, a region of semi-arid and arid landscapes in north-central Mexico enclosed by two parallel mountain ranges, the Sierra Madre Oriental and the Sierra Madre Occidental. The region has a limited agricultural base due to climate and elevation. The main industry within the Mesa del Norte is silver mining, depressed due to low international prices. The lack of work within the Mesa del Norte has caused many people to migrate to the United States or to northern Mexico where jobs can be found in maquiladoras.

• Migration to Monterrey's Maquiladoras

Monterrey, a city in northern Mexico, is the capital of the state of Nuevo León. Rapid population growth within Monterrey has made it the third largest city in Mexico after Mexico City and Guadalajara, with over 2.8 million people. One of the reasons for the growth of Monterrey lies in Mexico's economy. The rise of numerous manufacturing plants for foreign goods, called maquiladoras, has created a demand in Monterrey for large amounts of unskilled labor. People from the surrounding countryside are attracted to the city where they can earn wages more than double those offered in their home towns.

• The Future of the Mesa del Norte

With the passage of NAFTA, the Mesa del Norte region is expected to prosper. As wages in Monterrey rise, manufacturing plants are starting to decentralize to small towns on the Mesa del Norte. One town, Cedral, has seen the recent construction of two plants which manufacture womens' underwear. As new industry enters the region the economy will diversify, giving the people of the region more employment opportunities and stabilizing incomes. But NAFTA will also end price supports for many of the Mesa's agricultural goods. Many farmers are worried that they will be unable to compete with agricultural products from the United States.

• Spanish Conquest in Guatemala

The Spanish subjugation of the Central Highlands' Mayan Indians helped to create the settlement pattern seen in Antigua. Indian laborers were brought down from the highlands to work plantations for their new rulers, and this grid pattern helped to concentrate the population so that it could be more easily controlled. The fall of Spanish colonialism in Middle America allowed the Indians to return to their native lands and agricultural methods, which relied on a well-spread network of subsistence farms.

• Plantation System Imposed

After gaining independence in 1821, Guatemala became a capitalist state dependent upon plantation agriculture. Large estates dominated the rich land of the piedmont region, growing coffee, cocoa, sugar, and bananas. Often plantation land was obtained through illegal methods that forced out the local Indians, leaving them landless. The only way to earn a living was to work for a plantation during the harvest season. The rare photographs taken by Edward Muybridge capture the changes which have taken place in the coffee plantations and in beneficos where processing takes place. The village laborers shown in the Case Study work much as they did when photographed by Muybridge in 1876.

• Land Alienation Leads to Violence

The land required for the plantations to produce was confiscated from the Indians beginning in the 1870s. In less than 100 years, two-thirds of the arable land has been brought under the ownership of just 3 percent of Guatemala's farmers. The shrinking control of the land and the exploding population that land must support has led to bloody conflict.

• Military Occupation

Guatemala has been gripped by civil unrest for the past 30 years. As the military attempts to subdue a revolt by landless peasants, armed troops have moved into the Central Highlands, and the government has created a civil militia. Civil rights abuses by the Guatemalan government and military are common. The war has left more than 200,000 people dead and forced over 1 million people to flee their homes, becoming refugees within Guatemala and Mexico.

CASE STUDY CONNECTIONS

1. The intermingling of Spanish and Indian traditions and cultures characterizes these regions.
2. The physical environments of Mexico and Guatemala both have inaccessible highland regions and tropical climates.
3. The hacienda system of agriculture is employed in both countries. The plantation system supports a commodity-based export economy.
4. Great economic disparities exist within these countries, with land ownership mainly in the hands of the minority.
5. Economic development in Mexico and Guatemala is dependent upon outside investment and trade.
6. Domestically, both countries struggle with cultural and political stability.

GEOGRAPHERS AT WORK

The examination of the characteristics, distribution, and migration of human populations is a major field of study within geography.

In *Migration and Conquest,* Richard Jones of the University of Texas at San Antonio shares precedent-setting research assessing the volume, composition, and points of origin of undocumented Mexicans entering the U.S. In the course of identifying the source regions of illegal migrants, i.e., the Mesa del Norte, the economic dilemmas faced by these Mexicans becomes clear. Government policies in Mexico and the U.S. now strive to devise economic programs that will stem the tide of people moving from depressed regions, for example in agreements stipulated in the North American Free Trade Agreement which create job opportunities in maquiladora assembly plants in northern and north central Mexico.

The complex factors that either pull or push migrants within or between countries range from a search for employment to a desire to escape areas in a state of war, and the examination of these factors serves as the basis of geographical research on population movements of the present day.

TEST YOUR UNDERSTANDING

Questions

1. List three factors which encourage or push migration from the "Hollow Core."
2. Explain why NAFTA will be beneficial to Mexico. Explain why some people think that NAFTA will be detrimental to the Mexican economy.
3. Identify the "three conquests" of Guatemala.
4. Why have the peasant farmers of the highlands fought the Guatemalan government for over 30 years?

Map Exercises

The maps provided at the beginning of this unit can be used to complete this exercise. You may find extra copies of these maps helpful. Also, refer to the maps of Chapter 5, "Middle America: Collision of Cultures," from de Blij and Muller, 1994, *Geography: Realms, Regions, and Concepts: 7th Edition*, and any standard atlas to assist in this exercise.

1. Place the following names:

Mexico City	Monterrey	Mesa del Norte
Rio Grande River	Sierra Madre Oriental	Sierra Madre Occidental
Antigua	Guatemala City	Quiché District
Cuba	Barbados	Nicaragua
Costa Rica	Central Highlands	San Cristobal de las Casas

2. Create a thematic map showing population growth in the countries of Middle America. Use quintiles to group countries with similar growth rates.

CONSIDER THE DISCIPLINE

HISTORICAL GEOGRAPHY

Geographers study how landscape patterns evolve, and historical geography is one key to the interpretation of the turmoil of Middle America.

The Spanish came to **Guatemala** in the 16th century looking for wealth. They settled in the region near the piedmont of the Southern Coast and the Central Highlands, establishing **Antigua** as the seat of power for the Central American colony. When the city was destroyed in an earthquake in 1773, the capital was moved to Guatemala City (Ciudad de Guatemala). The Spaniards used Mayan labor to work the **plantations** at the lower elevations, growing cacao, indigo, and cotton. When

Spanish rule ended, many towns reverted from the tightly controlled **congregaciones,** dwellings spread out across the landscape established by the Spanish to control traditional land-use patterns. During the 19th century, new foreigners settled in the lower and mid-elevation region, stealing or buying up much of the land, thereby limiting the Indians' access to crop land and forcing the Mayans to once again work the plantations. After the 1944 revolution which overthrew the dictatorship of Jorge Ubico, a more democratic government was installed, slowly making reforms in **land use and ownership** and granting political freedoms to many leftist parties, such as the Communists. When President Jacobo Arbenz Guzman attempted to expropriate lands belonging to U.S. business interests, the United States backed an invasion that lead to the overthrow of Arbenz's government and installed a succession of **military rulers,** each supporting the wealthy, landed elite. In response, a **guerrilla rebel movement** arose, and bitter fighting began that has lasted for nearly 30 years, creating a **refugee population** of over 1 million people and leaving at least 200,000 people dead. In order to combat the rebels, the military has created local civil patrols of conscripted Mayan peasants who have been ordered to kill Indians found supporting the rebels.

The geographical forces which play a role in the **civil war,** namely the economic and social inequalities between the piedmont plantations and the highland agricultural system of the Guatemalan Indians, have yet to be addressed by the Guatemalan government. Until some type of land reform is implemented there is little hope for a peaceful resolution to the peasants' **struggle for equality.**

To further develop your understanding of this field of regional geography, refer to the Focus on a Systematic Field in Chapter 5, "Middle America: Collision of Cultures," from de Blij and Muller, 1994, *Geography: Realms, Regions, and Concepts: 7th Edition.*

CRITICAL VIEWING

Based on the Case Studies in Video Program 14: "Migration and Conquest" from *The Power of Place: World Regional Geography,* and the reading in Chapter 5, "Middle America: Collision of Cultures," from de Blij and Muller, 1994, *Geography: Realms, Regions, and Concepts: 7th Edition,* develop essays answering the following questions:

1. Many maquiladoras manufacture equipment for companies from the United States. Investigate the proportion of maquiladoras doing business with the U.S. and see if there have been any fluctuations in supply over the past 20 years.

2. Compare the land reform that has occurred in Guatemala with that in two other countries in Middle America. Investigate how land-use patterns have been affected by the different land reforms. What effects have land reform had on the populations of each country?

3. Discuss the cultural effects of subjugation of the Guatemalan people, the first by Spanish colonizers, second by international businesses, and finally by state terrorism designed to discourage support for subversives and rebels.

■ UNIT 7

South America: Continent of Contrast

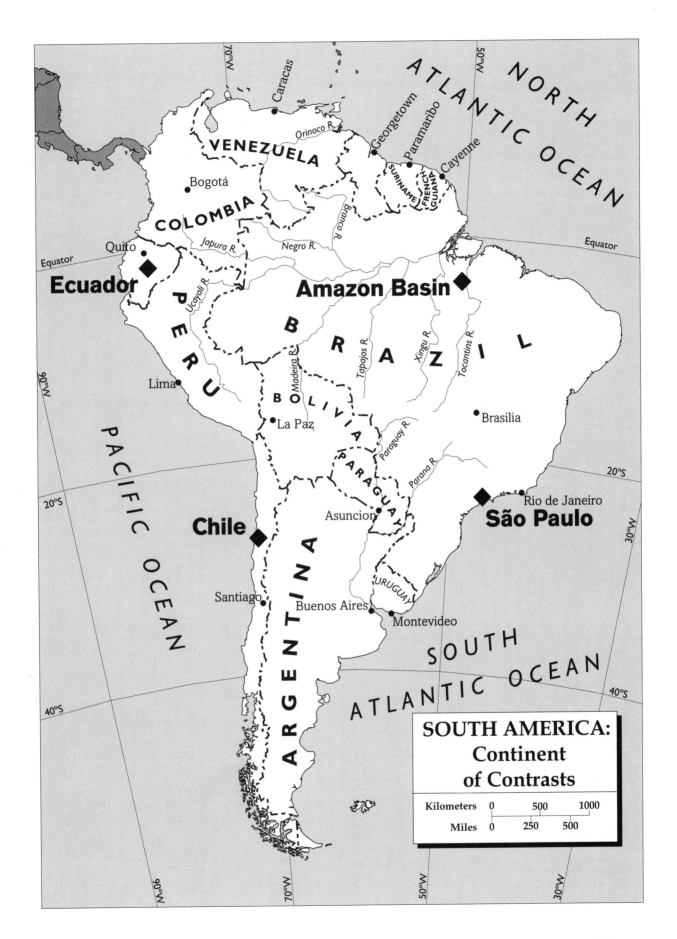

Caracas

ATLANTIC OCEAN

NORTH

VENEZUELA

Orinoco R.

Georgetown
Paramaribo
Cayenne

SURINAME

FRENCH GUIANA

Bogotá

COLOMBIA

Branco R.

Japura R.

Negro R.

Equator

Quito

Equator

◆ **Ecuador**

◆ **Amazon Basin**

P E R U

Ucayali R.

B R A Z I L

Tapajos R.

Xingu R.

Tocantins R.

Madeira R.

Lima

B O L I V I A

La Paz

Brasilia

Paraguay R.

PARAGUAY

Parana R.

20°S

◆ **Chile**

A R G E N T I N A

Asuncion

◆ **São Paulo**

Rio de Janeiro

30°W

PACIFIC OCEAN

20°S

Santiago

Buenos Aires

URUGUAY

Montevideo

SOUTH

40°S

ATLANTIC OCEAN

40°S

SOUTH AMERICA: Continent of Contrasts		
Kilometers 0	500	1000
Miles 0	250	500

90°W

70°W

50°W

30°W

▪ PROGRAM 15

Andes and Amazon

LESSONS IN GEOGRAPHY

At the end of Program 15, you should be able to:

1. Discuss how different maps of the same area may represent different information.
2. Develop a discussion about the meaning of sustainable development.
3. Identify the uses of satellites in finding and describing different locations on the earth's surface.
4. Understand the role of Brazil in the economic emergence of the South American continent.
5. Discuss how and why the hazards associated with volcanic eruptions in Ecuador are being studied.

OVERVIEW

Program 15 illustrates the power and fragility of the natural landscape in two South American countries. One Case Study examines how **humans** intrepret the dangers of their **environment**, and the next study explores how and why humans gauge their impact on the environment.

Ecuador: Valley of the Volcanoes chronicles the science and sociology of monitoring volcanic activity in the Andes. First, through tracking changes in an active **volcano,** then by mapping probable ash and lava flows, researcher Patti Mothes explores the ways in which **geographic tools** can facilitate better understanding of **natural hazards**.

A Second Chance for Amazonia? uncovers a new use of geographic **data collection** techniques for monitoring **sustainable development** policies in the **rainforests** of Northeast Brazil's Pará State. Here, research on the location of deforestation activity may hold clues to future forest management practices that **balance** agriculture, ranching, and forest land-use prospects.

STUDY RESOURCES

1. Video Program 15: "Andes and Amazon" from *The Power of Place: World Regional Geography.*
2. Unit 6 Map, "South America: Continent of Contrasts" from Latz and Gilbert, 1996, *The Power of Place: World Regional Geography Study Guide.*
3. Chapter 5, "South America: Continent of Contrasts" from de Blij and Muller, 1994, *Geography: Realms, Regions, and Concepts: 7th Edition.*

high risk
wont learn
due to: economy
possibilities
of survival
before it
happens

PREVIEWING QUESTIONS

1. What is the Global Positioning System?
2. Where is the Amazon River located?
3. What are the characteristics of the Ecuadorian Andes?

KEY THEMES

- **The Geographer's Tool**
- **The Andes**
- **Natural Hazards**
- **Global Positioning System**
- **Rainforest Sustainabililty**
- **Planning for Development**

DISCUSSION OF CASE STUDY THEMES

• The Geographer's Tool

Maps are geographic representations of people, places and environments in a spatial context. Geographers acquire, process and report data on maps to explain how the world is laid out. Researcher's examination of the geology, topography, and settlement patterns in Baños, Eucador is now graphically displayed in a non-technical map essential to educating the public about Tunguragua's eruptive potential.

• The Andes

The Andes Mountain Range stretches more than 4,500 miles along the entire western length of the South American continent. The range is home to more than 20 peaks, each over 20,000 feet high. Ecuador's central valley, or the Valley of the Volcanoes, is formed by two branches of the Andes. At least 200 volcanoes are in the chain, 30 of which are considered active. In 1985, a massive eruption in the Colombian Andes melted glaciers, triggering mudflows that killed more than 25,000 people.

• Natural Hazards

Vulcanism provides dramatic proof that humans live in a world ruled by physical processes. A sudden change, such as an eruption, can affect the land's capacity to support human activity and can produce disastrous results. The way humans perceive the risks associated with physical change and if and how they develop strategies to respond to them varies from denying any danger to launching public education campaigns. A devastating eruption 90 years ago did not stop the town of Baños from redeveloping. Inquiry into physical phenomena as well as human response may help mitigate future disasters.

• Global Positioning System

Through the advanced mathematics of measuring the sides and angles of triangles, a network of satellites can now precisely record the coordinates of any site on the earth's surface. A hand-held GPS device can calculate these triangulation measurements and display the latitude and longitude of a spot with an accuracy of a few feet to a few yards. Using GPS along with maps and air photos, field researchers can record land-use activities at exact locations.

smaller fields

• Rainforest Sustainability

New settlers in Amazonia are exploiting the area's natural resources. The principal economic activities which use these resources include logging, ranching, and agriculture. Conservationists argue that the rainforest ecosystem is too fragile to withstand the impact of this development. Geographer Christopher Uhl believes that a balance between conservation and development can be achieved. Using GPS in his research, Uhl concluded that the forest regenerates from its perimeter. Thus, to promote maximum regrowth, development should occur in island-type patterns. Replacing large clear-cuts with a series of smaller fields will promote a healthier ecosystem.

• Planning for Development

Geographic tools, such as GPS and air photos, are also being used to monitor compliance with a new land-use regulation allowing only 50 percent of the privately owned portion of the Brazilian rainforest to be developed. This statute, based on the sustainability ethos, acknowledges a compromise between preservation and economic progress.

Existing agricultural production is being redesigned for efficiency. An approached called intensification is now part of the development plan. If a given parcel of pasture can be better managed for its productivity and longevity, then fewer rainforests will be razed. This theory is practiced with techniques such as the intermixing of profitable crops and the use of higher-grade, less damaging fertilizers.

CASE STUDY CONNECTIONS

1. Both South American countries profiled are shaped by their physical attributes.
2. Both the Andes Mountains and the Amazon Basin are unique geographic features on the earth's surface which attract international researchers.
3. Maps and geographic data collection are used for planning in both places.
4. Satellites and computers improve and expand spatial analysis of remote locations.
5. How humans interact with the environment can be represented by geographic tools. In Ecuador, a map is being used to show escape routes in case of volcanic eruption, and GPS data are used to track deforestation in Brazil.
6. Environmental change due to natural and human activity is a common denominator in each Case Study.

GEOGRAPHERS AT WORK

Field work and respect for empirical research are the hallmarks of the geographic approach to studying the environment. The research agenda of Patty Mothes illustrates the field's long-standing environmental concerns through study of the ways people make decisions relative to environmental uncertainty and risk in the Ecuadorian region of the Andes Mountain Range.

Through field work, Mothes is collecting information that will be used to model possible scenarios describing how the eruption of Mt. Tunguragua will spew gas, lava and ash on the nearby community of Baños. The applicability of natural hazards research to public policy is clear cut in this case, as an attempt is made to devise evacuation plans to protect people dwelling in the most dangerous areas adjacent to the mountain.

The geographer's attempt to synthesize information about the local physical environment and humans' perception of it is as old as the art of cartography. Since ancient times, geographers have compiled elaborate maps detailing the location and distribution of people and natural phenomena. Field work, however, is also as modern as 20th century resource management theory, similar in its imperative to examine the decision-making and environmental assessment components that comprise contemporary natural hazards research.

TEST YOUR UNDERSTANDING

Questions

1. What is the practical application of the geographic research in Baños?
2. List three natural hazards in South America.
3. How do geographers use modern computer technology to better understand the environment.
4. Why has settlement in the Amazon Basin grown dramatically in the past 15 years?

Map Exercises

The maps provided at the beginning of this unit can be used to complete this exercise. You may find extra copies of these maps helpful. Also, refer to the maps of Chapter 6, "South America: Continent of Contrasts," including its vignette focusing on Brazil in de Blij and Muller, 1994, *Geography: Realms, Regions, and Concepts: 7th Edition*, and any standard atlas to assist in this exercise.

1. Place the following names:

Andes Mountain Range	Amazon River	Ecuador
Quito	Baños	Brazil
Pará State	Amazonia	Paragominas
Belém	Bolivia	The Equator

2. Trace the Amazon River on a map of South America. Map the zone of its drainage network. Indicate the extent of the Brazilian rainforest.

CONSIDER THE DISCIPLINE

THE GEOGRAPHY OF SPATIAL ECONOMY: PRIMARY ACTIVITIES

Both national and international forces are reshaping the economic geography of the Amazon, the single largest and most important **rainforest** in the world.

Located primarily in the country of Brazil, the Amazon is a **developing region** whose population has nearly tripled to 15 million since 1980. Ongoing development projects include hydro-electric power, timber harvesting, commercial agriculture, ranching, and the creation of huge mining complexes which attract hundreds of thousands of settlers who have chosen to migrate in much the same way pioneers of westward expansion did in the U.S. during the 19th century. Government

Belem

programs creating transportation networks, highway and rail systems, and provision of inexpensive land deeds, are important efforts to raise the standard of living in Brazil, to expand the national economy, and to reduce the crushing poverty afflicting so much of the population. The most remarkable forces altering the Amazon are the **international economic pressures** associated with the commercial harvesting of the rainforest due to global demand for housing, paper, and furniture. The rapid deforestation that is occcuring, approximately 20,000 to 30,000 square miles per year in the early 1990s, has become an international concern. Scientists have concluded that these rainforests are the richest and most **diversified ecological habitat** on the face of the earth. Amazonia, and other tropical rainforests in the world, contain an estimated half of all plant and animal species, valuable medicinal, food, and other products, and countless plant-based drugs for fighting disease. By far the most serious consequence of rainforest destruction is the increasingly widespread scientific view that these fragile environmental zones play a disproportionately large role in the maintenance of the climate patterns that nourish life across the face of the earth. With the disappearance of these rainforests may come an increase in average global temperatures and environmental catastrophes as icecaps melt and coastal areas supporting the majority of the world's population are threatened by rising sea levels.

The Amazonian region of South America is distinguished by its natural environment as well as the spatial extent of the economic forces which are now transforming the region. Both domestic and international factors alter Amazonia, and the resulting changes in agricultural, forest, and mining land uses and in human settlement patterns have consequences that extend beyond the physical and economic region of the Amazon Basin. **Geographical analysis** helps to identify places like Amazonia on the face of the earth that are particularly susceptible to economic development, and when the area is large, fragile, and endowed with extremely valuable resources, as in the case of the Amazon rainforest, the consequences of local activity can be global in scope. In assessing the future prospects for the Amazon, scientific analysis indicates that the agents of change transforming the area are both national and international, yet the formulation of policies to solve the problems associated with rainforest exploitation hinge on the recognition of a geographic imperative: agreements about **site-level management** of fragile resources must support both the needs of people who dwell there and the needs of a larger international community which may be adversely affected by rainforest depletion.

To further develop your understanding of this subfield of regional geography and its relationship to Brazil's development, refer to the Focus on a Systematic Field and the vignette "Emerging Brazil: Potentials and Problems" in Chapter 6, "South America: Continent of Contrasts," from de Blij and Muller, 1994, *Geography: Realms, Regions, and Concepts: 7th Edition*.

CRITICAL VIEWING

Based on the Case Studies in Video Program 15: "Andes and Amazon" from *The Power of Place: World Regional Geography*, and the reading in Chapter 6, "South America: Continent of Contrasts," from de Blij and Muller, 1994, *Geography: Realms, Regions, and Concepts: 7th Edition*, develop essays answering the following questions:

1. Geographer Christopher Uhl stated that a practical approach to Brazilian rainforest depletion recognizes that development is inevitable. Propose an alternative development plan that balances the economic needs of the large

Andes and Amazon

number of people moving into the area, international environmental concerns about rainforest preservation, and the integrity of the rainforest ecosystem.

2. What practical applications are now possible for geographic research using GPS technology? What is the link between the technological possibilities for data collection and their application to public policy?

3. List three natural hazards that could effect the region where you live. What is the public perception of the level of danger of these hazards?

Sao Paulo
new arrivals in
live in
rings
around
city

■ PROGRAM 16

Accelerating Growth

LESSONS IN GEOGRAPHY

At the end of Program 16, you should be able to:

1. Describe the settlement patterns of South America's mega-city, São Paulo, the world's third largest city.

2. Review the major migrations that have given São Paulo a variety of ethnic neighborhoods.

3. Examine the development of São Paulo's shantytowns and explain why they have grown at the city's edge.

4. Discuss how neighborhood groups in the shantytowns are trying to integrate their communities into Greater São Paulo.

5. Understand the historical influence of European culture on modern Chilean society.

6. Trace the source of Chile's new wealth, the export-oriented economy that has linked Chilean businesses to countries across the globe.

7. Examine the changes that export agriculture has brought to the northern grape-growing region of Chile's Limari Valley.

8. Explain the unique geography of Chile, which allows the country to grow a diverse range of agricultural products for the world economy.

Chile –
best
economy

OVERVIEW

This program, the second of two that explore the realm of South America, looks at two of the continent's most rapidly growing places: Chile, where the economy is the most robust on the continent, and São Paulo, where the recent influx of migrants from Northeast Brazil is the latest in a long history of immigration that has created the third largest urban center in the world.

In *São Paulo: The Outer Rim* the video shows how a small Portuguese settlement grew into the **world's third largest city** and the largest city on the South American continent. This city in southeastern Brazil has long been a leading center for Brazil's manufacturing sector, as well as for heavy industry, financial organizations, and petroleum refining. In addition, the São Paulo region leads Brazil's coffee production, producing more than one-third of the country's total output.

The recent growth of São Paulo into a mega-city is a part of a trend of **urbanization** across South America. Over 70 percent of the continent's population now live in urban areas, 20 million people in São Paulo alone. Looking across the cityscape, the scale of the city is revealed. **Shanties** encircle the city for miles and miles, full of new immigrants from Northeast Brazil. These new settlers build with whatever materials they are able to find and afford, but they receive little help from the cosmopolitan heart of the city. Most of the homes in the shanty areas are built by squatters with no legal ownership of the land, and no public services are provided by the city.

The second Case Study, *Chile: Pacific Rim Player,* explores Chile's unique geography and the role that this geography has played in the country's shift away from its **traditional economy** and European connections. The country's history is rooted in the Indian cultures of the Andes Mountains, but little of this heritage has been reflected since the Spanish penetrated of Chile in 1536. Nearly 300 years of Spanish rule has left its imprint on the architecture and demography of Santiago, Chile's capital and primary city. Recently Chile has worked toward integrating itself into both the South American Realm and countries of the **Pacific Rim**, building a dynamic, growing economy on the strength of its agricultural and forestry sectors. **Global interaction** is once again changing the city, and the country along with it.

STUDY RESOURCES

1. Video Program 16: "Accelerating Growth" from *The Power of Place: World Regional Geography.*
2. Unit 7 Map, "South America: Continent of Contrasts" from Latz and Gilbert, 1996, *The Power of Place: World Regional Geography Study Guide.*
3. Chapter 6, "South America: Continent of Contrasts" from de Blij and Muller, 1994, *Geography: Realms, Regions, and Concepts: 7th Edition.*

PREVIEWING QUESTIONS

1. Where is Chile located on the Pacific Rim?
2. Describe how physical geography influences agricultural production.
3. What are the characteristics of a mega-city?
4. In what ways does ethnic migration contribute to urban settlement patterns?
5. How do shanty communities evolve and where are they typically located?

KEY THEMES

• **The Mega-City of São Paulo**
• **A City of Immigrants**
• **Growing Shantytowns**
• **Land-Use Patterns of São Paulo**
• **Exploiting Geographic Advantages to Fuel the Chilean Economy**
• **Changing Economic Patterns in the Limari Valley**
• **Social Impacts of Change**
• **Chile's Changing Export Patterns**

DISCUSSION OF CASE STUDY THEMES

• The Mega-City of São Paulo

São Paulo is large by any standard, with 20 million residents spread out on an expanse of land that stretches more than 60 miles (100 km) from the city center. São Paulo ranks as the third largest city in the world, behind Tokyo and Mexico City, and, if current projections hold true, may be the largest by 2010.

• São Paulo – City of Immigrants

São Paulo's history is filled with stories of immigrants carving out their niche within the city, starting with the first inhabitants, the Portuguese. For 300 years, São Paulo's population slowly grew to 2 million people. When slavery was abolished throughout the country in 1888, São Paulo experienced its next rush of immigrants as blacks from the north streamed down to the city looking for employment. During this time, a major influx of foreign immigrants, mostly Italians, came to São Paulo, and by the 1950s more than 5 million had arrived. The Italians settled in the neighborhood of Bixiga, sharing it with blacks from North Brazil. At that time, the neighborhood was on the fringe of the city and its people were very poor. As the city has grown, Bixiga has become a prosperous neighborhood and an important part of the city's core. Japanese immigration began early in the 20th century and centered around the Liberdade region. Today São Paulo boasts the largest population of Japanese outside of Japan.

• Growing Shantytowns

By 1960 the city had reached a population of almost 13 million, but still more people were on the way. The newest immigrants began to construct homes and neighborhoods on the periphery of the city in a process called "self-construction." The resulting unplanned, sporadic, disorganized layout of the shanties has caused the city to grow into a metropolis with a 60-mile radius. Most shanty neighborhoods are not officially recognized by the city, but as the cardboard and scrap metal homes are replaced with more permanent structures, there is increasing pressure to provide them with city services: public transportation, sewer and water lines, schools, and electricity. With luck and hard work, the people in shanty neighborhoods hope to build and improve their lives by linking themselves, as previous immigrants did, with the thriving city in the distance.

• Land-Use Patterns of São Paulo

São Paulo has a vibrant core region, with a triangular pattern carved by the Portuguese that still can be seen today. As new residents move into the city they create neighborhoods on its edge, slowly expanding the city and integrating themselves into the city service district. This pattern in present-day São Paulo continues in the new shanty neighborhoods that arise, well-removed from the core area and disconnected from city services. As the shanties grow, only some are integrated into the city, while most still suffer sub-standard conditions.

Handwritten margin notes (top): Most exports · Climate free growing season (opposite of NA) Nov-March

Handwritten margin note (left, top): rural land use

• Exploiting Geographic Advantages to Fuel the Chilean Economy

The diversity of climates found within Chile, the result of the country's physical geography and shape, is a key to its explosive economic growth. The country stretches 2,700 miles (4,500 km) north-south along the western shore of South America, passing through a range of latitudes. East-to-west, the country averages only 110 miles (177 km) but contains three distinct physiographic features: the Andes Mountains, the Pampa Central, and the coastal ranges. The resulting north-to-south climatic variability allows for the production of a great variety of crops, from fruits to timber products. Beyond table grapes and timber, the mountains are rich with minerals such as copper, lithium, silver, and gold, and the fisheries in the Pacific Ocean are well endowed by the nutrient-rich Humboldt Current, also called the Peru Current. In addition, Chile benefits from its position in the Southern Hemisphere, where the growing season falls between November and March, allowing it to serve produce markets in the Northern Hemisphere during the winter months when off-season local production cannot meet demand.

• Changing Economic Patterns in the Limari Valley

Handwritten margin note (left): small farmers + local needs are hurt by large grape export industry

The growth of the table grape industry in the Limari Valley has had a substantial impact on the agricultural land-use and settlement patterns of the valley. As more land is put into grape production, small farmers are squeezed off their farms and forced to seek employment at large vineyards. Further consequences arise when land fulfilling local needs is taken out of production, forcing residents to rely on food from outside sources.

• The Changing Social Geography of the Limari Valley

Handwritten margin note (left): migrants; women; home

As land-use practices change, the social geography of the Limari Valley is also altered. The harvest, a labor-intensive process, draws large numbers of migrant workers from the countryside and cities. With this influx of people, which more than quadruples the population, come urban problems such as crime, drug use and alcoholism. Change is also occurring in the home. The new vineyards have presented employment opportunities for the women of the valley. Yet, as more women enter the workforce, families spend less time together, challenging traditional expectations.

• Chile's Changing Trade Patterns

Handwritten margin note (left): wood · US Germany Japan Brazil

The cool, wet climate of southern Chile is perfect for the Radiata Pine, imported from North America over 20 years ago. In Chile the trees are able to mature in just 20 years, in contrast to 30 years in North America, and the tall, straight trees produce sturdy lumber. Other important products in Chile's export-oriented economy include paper, chemicals and petroleum. As the Chilean economy becomes increasingly dynamic in the world market, a shift in export destination is occurring. For many years, the United States and Germany were the country's chief trading partners. As Chile begins to reap the rewards of its unique geographic advantages, new markets are sprouting across the Pacific Ocean, and today Japan, along with the United States and Germany, ranks as one of Chile's leading trade partners. Across the Andes, Brazil has also climbed high on this list of trading partners. Chile's emphasis on an export-oriented economy has allowed the country to aggressively seek out new markets for its products. Some observers now refer to Chile as the next "economic tiger" on the Pacific Rim.

Handwritten margin note (bottom): forest products in southern coastal range

small farmers hurt by exporters

danger if only extractive part of process — not producers — need to use skill.

CASE STUDY CONNECTIONS

1. Each country's population, history, and economy are tied to European immigrants, especially those from Portugal and Spain.

2. Chile has one of the most robust economies in South America; Brazil aspires to boost its economy to such a level.

3. Forces of global economic integration are shaping agriculture and mining, the primary economic sectors of both countries.

4. Land-use patterns, both urban and rural, are undergoing rapid transformations in both countries.

5. Changes in domestic economies have led to the rapid growth of poor populations concentrated in particular rural and urban geographic areas.

6. Subsistence agriculture is being replaced by commercial agriculture in each country, leading to significant rural-to-urban migration streams, often by diverse ethnic groups.

TEST YOUR UNDERSTANDING

Questions

1. Discuss the causes of the growth of the shanty neighborhoods surrounding São Paulo.

2. Why is the Limari Valley well suited to grow table grapes for export?

3. Besides table grapes, list three products resulting from Chile's unique geography which have been able to penetrate the world economy.

Map Exercises

The maps provided at the beginning of this unit can be used to complete this exercise. You may find extra copies of these maps helpful. Also, refer to the maps of Chapter 6, "South America: Continent of Contrasts," from de Blij and Muller, 1994, *Geography: Realms, Regions, and Concepts: 7th Edition*, and any standard atlas to assist in this exercise.

1. Place the following names:

Brazil	São Paulo	Rio de Janeiro
Brasilia	Chile	Limari Valley
Atacama Desert	Andes Mountains	Humboldt Current
Uruguay	Argentina	Great Escarpment
Santiago	Northeastern Brazil	Chilean Coastal Range

1994 Japan deal w/ Chile

Accelerating Growth

2. Using data documenting the changing patterns of trade over the last two decades, draw a map detailing the current trade patterns of Chile. Use arrows to indicate the percentage of trade flow among Chile and its four largest trade partners.

CONSIDER THE DISCIPLINE

THE GEOGRAPHY OF SPATIAL ECONOMY: PRIMARY ACTIVITIES

As Chile has begun to exploit its geographic advantages, the economy has been able to **diversify** away from another of its traditional **primary extractive products,** copper. In the 1970s copper exports represented over 70 percent of the country's production, but today, the role of fish, forest products, fruits and vegetables has expanded, reducing cooper's proportion in **international trade** to less than 40 percent. One place in which these changes can readily be seen is the valley of the Limari River in the Norte Chico region of Chile. Ten years ago, the steep hillsides of the valley were barren, but due to the unique micro-climate, table grapes become ripe just in time for the Christmas holiday season in North America. With help from **irrigation** works, the Limari Valley has seen tremendous expansion of land under production in the past ten years. Large vineyards now dominate the valley, changing the cultural as well as the physical landscape.

As land has been converted from small-scale, **subsistence agriculture** designed for local consumption to large-scale **commercial agriculture** for international markets, people have been pushed off their own land and forced to work for the large farms. During the harvest season, Chanaral Alto swells from 3,000 to 14,000 people, and some migrant workers bring urban problems, such as increased criminal behavior and drug and alcohol abuse, with them. These new large farms have also changed the role of women in the region, giving them more employment opportunities but often disrupting family life.

To further develop your understanding of this subfield of regional geography, refer to the Focus on a Systematic Field in Chapter 6, "South America: Continent of Contrasts," from de Blij and Muller, 1994, *Geography: Realms, Regions, and Concepts: 7th Edition.*

CRITICAL VIEWING

Based on the Case Studies in Video Program 16: "Accelerating Growth" from *The Power of Place: World Regional Geography,* and the reading in Chapter 6, "South America: Continent of Contrasts," from de Blij and Muller, 1994, *Geography: Realms, Regions, and Concepts: 7th Edition,* develop essays answering the following questions:

1. Compare the growth of São Paulo with that of another mega-city, such as New York City or Tokyo. Investigate migration patterns and birth rates to see how the two cities have expanded. Examine and compare the patterns of physical expansion of the cities.

2. Compare the changing land-use patterns of the Limari Valley in Chile with land use in Guatemala. What similarities are there? How does the current pattern in Chile differ from the large haciendas in Guatemala?

■ UNIT 8

North Africa/Southwest Asia:
The Challenge of Islam

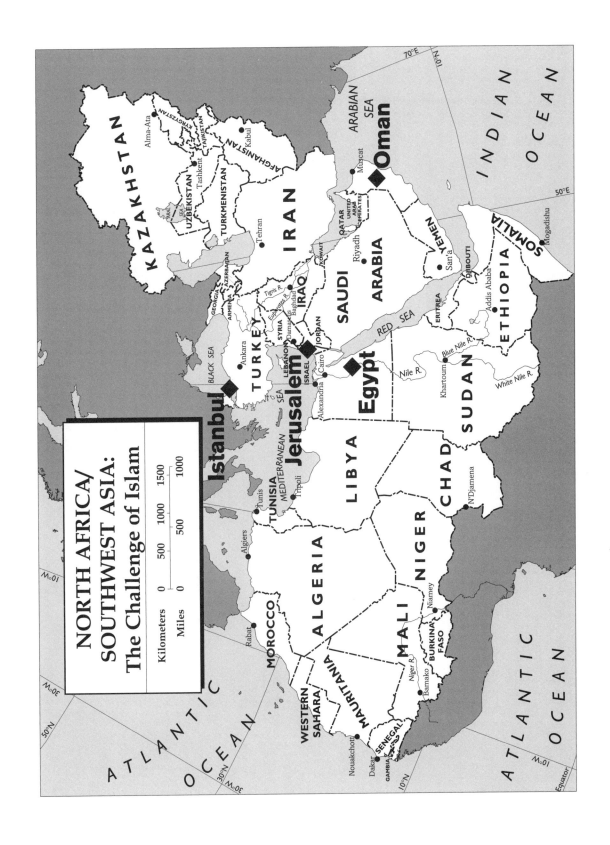

NORTH AFRICA/
SOUTHWEST ASIA:
The Challenge of Islam

Kilometers

0   500   1000   1500

0   500   1000

Miles

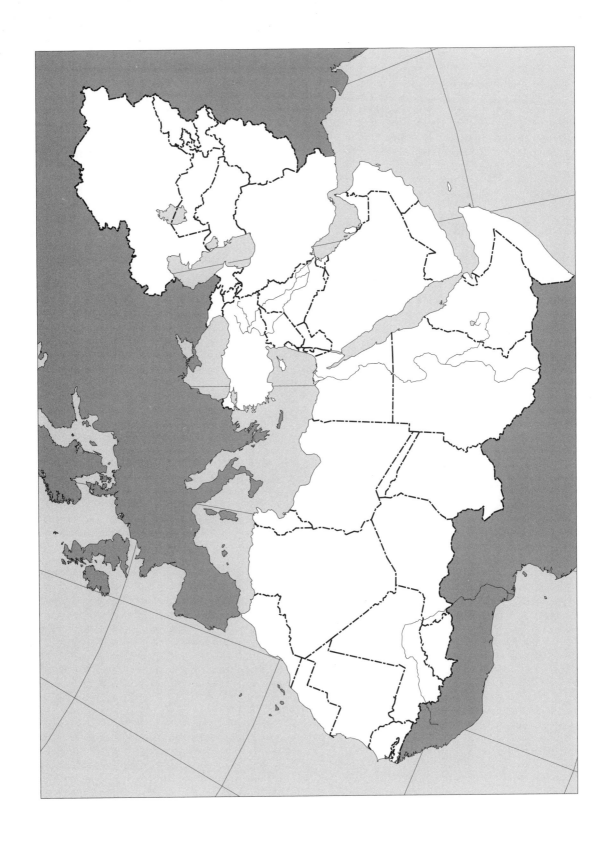

▪ PROGRAM 17

Sacred Space Under Siege?

LESSONS IN GEOGRAPHY

At the end of Program 17, you should be able to:

1. Explain why places such as the holy city of Jerusalem are important as symbols for unifying or fragmenting society.

2. Analyze how cooperation and conflict influence the development and control of social, political, and economic entities surrounding Jerusalem.

3. Explain the mass migration into the city of Istanbul.

4. Evaluate the impact of large-scale rural-to-urban migration on Istanbul.

5. Appreciate the importance of Islam in shaping fundamentalist attitudes within Turkey.

OVERVIEW

Program 17 explores spatial variations of **religious practice** in Jerusalem and examines the gaps between rich and poor, secular and **fundamentalist**, in Istanbul.

In *Jerusalem: Sacred Space Under Siege?* the three major monotheistic religions of the world share the same sacred ground in the Old City, but without peace. The boundaries carving up this city and the rest of the **Holy Lands** were wrought from warfare. But the issues are far from settled.

Jerusalem means "city of peace," but Israeli soldiers are always on guard. The entire city is divided by the **Green Line**, which was the Israeli eastern boundary before it captured East Jerusalem and the West Bank in the Six Day War of 1967. Part of the land captured in this war includes the Old City and its significant religious sites. **Palestinians** believe that Israeli control beyond the Green Line and settlements in the West Bank occupy the land of their would-be state.

At the junction of Europe and Asia, western and eastern influences collide in *Istanbul: Fundamental Change.* Critics of **modernization** in Turkey's cities see secular democratic and economic reforms as corrupting fundamental Islamic ways of life. This Case Study explores the challenges of the rural-to-urban **migration** that brings secular Turks and fundamentalist Muslims head-to-head in Turkey's largest city.

The video contrasts Istanbul's oriental atmosphere with evidence of its Westernization, examining the wealth in some of its districts and the poverty in others, and shows how its secular society is giving rise to fundamentalism in its politics.

STUDY RESOURCES

1. Video Program 17: "Sacred Space under Siege?" from *The Power of Place: World Regional Geography*.

2. Unit 8 Map, "North Africa/Southwest Asia: The Challenge of Islam" from Latz and Gilbert, 1996, *The Power of Place: World Regional Geography Study Guide.*

3. Chapter 7, "North Africa/Southwest Asia: The Challenge of Islam" from de Blij and Muller, 1994, *Geography: Realms, Regions, and Concepts: 7th Edition.*

PREVIEWING QUESTIONS

1. Where is the Old City within Jerusalem? Can you name the Quarters within the Old City?

2. Which religions worship at the Western Wall, the Temple Mount, and the Church of the Holy Sepulcher?

3. What is the present-day name of Constantinople? When and why did the name of the city change?

4. Is Turkey considered to be in the Arab World, the Islamic World, or both?

KEY THEMES

- **History of Israel**
- **Urban Geography of Jerusalem**
- **Territorial Conflict in the Holy Lands**
- **The Separation of Church and State in Turkey**
- **Migration to Istanbul**
- **Fundamentalism in Politics**

DISCUSSION OF CASE STUDY THEMES

• History of Israel

Long before the creation of modern State of Israel, Judaism, Christianity, and Islam vied for control of the Holy Land. Bible stories relate conflict over the lands of the Israelites before Christ. After Jews were exiled from Jerusalem to Babylon in 586 B.C., they returned to rebuild their Temple, but in 30 A.D. they were defeated by the Romans. After 300 years of Christian Byzantine rule, and just six years after the death of the Prophet Mohammed in 632, Muslims conquered Jerusalem and built the Dome of the Rock shrine over the Foundation Stone of the Temple where, in Islamic belief, Mohammed ascended to heaven. The Early Arab Period continued until Christian solders mounted the Crusades of 1099 and slaughtered many of the Jews and Muslims of Jerusalem. Another Muslim era began 88 years after the Crusades and lasted until the Ottoman Turks settled in the city in 1517. Near the end of World War I, British troops drove the Ottoman Turks from the city. The League of Nations then gave Britain a mandate to govern Palestine, including Jerusalem, in 1922. Toward the end of British Rule, the newly constituted United Nations called for the partition of Palestine and the internationalization of Jerusalem. Arabs rejected this plan, war broke out and the State of Israel declared independence on May 14, 1948.

• Urban Geography of Jerusalem

The walled Old City on the eastern side of Jerusalem's Green Line is divided into religious and ethnic quarters. The Jewish and Muslim Quarters are located alongside the Western Wall remnant of the destroyed First and Second Temples' and the Dome of the Rock at the Temple Mount. Only 1,000 feet to the west, the Christian and Armenian Quarters claim sacred ground. The traditional site of Christ's crucifixion, burial, and resurrection sits in the middle of the Christian Quarter at the Church of the Holy Sepulcher. Today, the church is physically and spiritually divided among Roman Catholics, Greek Orthodox, Armenians, Syrians, and even Ethiopian monks and nuns, denied part of this sanctuary, have lived on the roof since the mid-1800s to press their claim for shared ownership.

• Territorial Conflict in the Holy Lands

At the end of Israel's War of Independence, Jerusalem's western side was controlled by Israel; its eastern side, including the Old City, by Jordan. After the Six Day War of 1967, Jerusalem was taken by Israel, but Muslims were granted control of the Temple Mount in the Old City of Jerusalem. In 1980, a special law passed by the Israeli parliament reasserting its capital city in Jerusalem was refuted by Palestinians who also laid claim to the same capital for their proposed state. By 1993, Israeli and Palestinian peace negotiators signed a Declaration of Principles that outlined the terms of Palestinian self-government in the Gaza Strip and Jericho, but the most sensitive issue — the fate of Jerusalem — will not be discussed until the final stages of negotiations.

• The Separation of Church and State in Turkey

In the 1920s, after the last remnants of the Ottoman Empire had disappeared in World War I, Kemal Atatürk built the modern state of Turkey. One of President Atatürk's major political achievements was the complete formal separation of Church and State. He reduced the power of Islam by abolishing the caliphate (the office of supreme ruler of a Muslim state), substituting Roman for Arabic lettering and purging Turkish of Arabic words. He declared that religion should be a matter of individual choice and conviction only.

In 1934, he passed a law requiring all Turks to use surnames in the Western style and himself took the name Atatürk — father of the Turks.

• Migration to Istanbul   *when Europe & Arab World met*

Europe & Asia

Another of Atatürk's achievements was economic modernization, most clearly evident in urban areas. As a result, tremendous pull factors encouraged rural-to-urban migration, and by the 1950s and 1960s many villages were completely abandoned. Migration to big cities such as Istanbul is also political. An internal war against the traditionally nomadic Kurds in the southeast, over their demand for self-government, has contributed to this migration pattern by driving refugees out of the countryside and into the cities.

The latest census in 1990 counted 7 million inhabitants in Istanbul, but now the local geographer interviewed assumes that the city houses more than 10 million. Immigration is estimated at 300,000 to 400,000 people per year. The result is massive growth.

• Fundamentalism in Politics

After arriving in Istanbul, immigrants vie for jobs and housing on the outskirts of the city. Many residences are self-built around once-small villages, producing shanty town conditions. In such overgrown, under-serviced settlements, the fundamentalist

Sacred Space Under Siege?   153

Welfare Party is gaining popularity. The movement, which observes only Islamic laws and divine authority, holds an appeal for those who are not well served or even suffer under the modern western-style economy and the secular government.

CASE STUDY CONNECTIONS

1. Religious beliefs are at the root of social, political and, ultimately, territorial conflict.
2. Different perspectives on location, settlement, and public policy in Jerusalem and Istanbul are voiced by Jews, Muslims, and Christians.
3. Both Jerusalem and Istanbul face two possibilities: further fragmentation or unification amongst competing interests.
4. The Palestinians and the Kurds are nations without states.
5. Istanbul and Jerusalem both continue to draw large refugee populations.
6. Islamic fundamentalist beliefs are opposed to the status quo in both locations.

GEOGRAPHERS AT WORK

Differences in cultural traits, notably religion, language, and ethnicity, can lead to political problems at the urban and even the national scale. Professor Kamal Abdul Fattah, of Birzeith University, studies such issues of cultural and political geography in his research on Jerusalem and prospects for a Palestinian state.

Arabs, Christians, and Jews have laid claim to Jerusalem for thousands of years. As a result, the city's cultural geography is particularly complex, among the most complex in the world. When there is a long and protracted tradition of conflict among different cultural groups, one of the consequences is often an impermanence of political boundaries and therefore a degree of political instability.

The 20th century is replete with examples of efforts by diverse and contesting groups of people to divide the earth's surface. War is one manifestation of these groups' deeply held desires to organize and divide areas according to their particular values and perceptions. At present, the earth is primarily politically divided by state sovereignty, and it is this manner of division which continues to challenge Israelis and Palestinians in the disputes surrounding Jerusalem.

TEST YOUR UNDERSTANDING

Questions

1. Which part of Jerusalem is under contested ownership?
2. How long has control been contested?
3. How are religious convictions reflected in the landscapes of Jerusalem? In Istanbul?

Map Exercises

The maps provided at the beginning of this unit can be used to complete this exercise. You may find extra copies of these maps helpful. Also, refer to the maps of Chapter 7, "North Africa/Southwest Asia: The Challenge of Islam," from de Blij and Muller, 1994, *Geography: Realms, Regions, and Concepts: 7th Edition*, and any standard atlas to assist in this exercise.

1. Place the following names:

Istanbul	Kurdistan	Ankara
Aegean Sea	The Bosphorus	Black Sea
Euphrates River	Atatürk Dam	Iraq
Mediterranean Sea	Israel	Jerusalem
The Old City	West Bank	Jordan

2. Map the countries bordering Turkey. Show why Turkey has been considered a land bridge for trade and cultures.

CONSIDER THE DISCIPLINE

CULTURAL LANDSCAPE OF JERUSALEM

Cultural values and priorities shape the nature and extent of human activity across the face of the earth. As a field of study, **cultural geography** explores the imprint of human activity that grants distinctive identity to the **landscape** found in particular places. The cultural landscape of the Old City of Jerusalem, the spiritual hub of several religions and the location of their sacred sites for 4,000 years, is a case in point.

In Jerusalem, one finds a distinctive cultural landscape represented in the city form — the architecture of Jewish, Islamic, and Christian buildings. The **competition** among religions for **territorial control** also results in cultural **imprints** on the neighborhoods of the Old City, the distinctive Muslim, Jewish, and Christian Quarters. The **intersection of politics and culture** is further illustrated by the way values found in religious systems serve to imbue particular places with sacred meaning, e.g., the Western Wall and the Dome of the Rock. Who should exercise sovereignty over sites sanctified as holy represents the most difficult conflict in this part of the world. This fact serves as a backdrop as to why Judaism, a fundamentally ethnic religion, exhibits values characteristic of a religious system whose institutions and practices closely link that ethnic group to the land of its origin. This is the religious context in which the sacred meaning is attached to the very land which comprises the State of Israel, the West Bank, and the Old City of Jerusalem.

The overlap of religious landscapes in Jerusalem is a reflection of the cultures associated with this urban center, and the resulting **cultural mosaic** is elaborate and meaningful to the people who dwell there.

To further develop your understanding of cultural geography, refer to the Focus on a Systematic Field in Chapter 7, "North Africa/Southwest Asia: The Challenge of Islam," from de Blij and Muller, 1994, *Geography: Realms, Regions, and Concepts: 7th Edition*.

CRITICAL VIEWING

Based on the Case Studies in Video Program 17: "Sacred Space under Siege?" from *The Power of Place: World Regional Geography,* and the reading in Chapter 7, "North Africa/Southwest Asia: The Challenge of Islam," from de Blij and Muller, 1994, *Geography: Realms, Regions, and Concepts: 7th Edition,* develop essays answering the following questions:

1. How do differing perceptions of the distribution of territory surrounding and encompassing Jerusalem cause cultural conflict and fragmentation?

2. Explain how the extent of modernization and its geographic impact on Istanbul has encouraged the rise of fundamentalism. Give specific examples.

3. Based on the principles of relocation diffusion, how will Kurdish refugees influence the spread of the Welfare Party in Istanbul?

▪ PROGRAM 18

Population, Food Supply, and Energy Development

LESSONS IN GEOGRAPHY

At the end of Program 18, you should be able to:
1. Explain the population distribution along the Nile and contrast it to the rest of Egypt.
2. Understand how the Aswan Dam complex changed the landscape of Egypt.
3. Discuss how the magnitude of the population in Cairo endangers the prime agricultural land of the Nile Delta.
4. Develop a definition of Omanization.
5. Describe the role of oil in the past and future economies of Oman.
6. List the policies of modernization of Oman.

OVERVIEW

Program 18 visits Egypt and Oman in the North Africa/Southwest Asia Realm. Both Case Studies examine the interaction between humans and natural resources.

An exploding **population growth rate** and a shrinking **arable land** base is the dilemma presented in *Egypt: Population Overload*. A soil scientist recounts the modern history of the control of the Nile River. While the Egyptian desert adjacent to the Nile has bloomed, **agricultural production** still lags far behind demand. Facing **exponential population growth** of nearly 3 percent, Egypt already must import 50 percent of its food supply. The government's search for **habitable land** to provide housing and services directly threatens the arable land needed to feed its citizens.

A newly modernized country tries to **diversify** its economy in *Oman: Looking Beyond Oil*. Nomadic in nature until the discovery of oil in the 1960s, Oman has modernized atop energy resources in the past 30 years. Seventy-two percent of the Omani economy is now based on oil. To avoid **overdependence** on one extractive commodity, Sultan Qaboos has enacted a **national plan** of domestic worker training, development of new economic sectors, and replacement of foreign guest workers called **Omanization.**

STUDY RESOURCES

1. Video Program 18: "Population, Food Supply, and Energy Development" from *The Power of Place: World Regional Geography*.

2. Unit 8 Map, "North Africa/Southwest Asia: The Challenge of Islam" from Latz and Gilbert, 1996, *The Power of Place: World Regional Geography Study Guide.*

3. Chapter 7, "North Africa/Southwest Asia: The Challenge of Islam" from de Blij and Muller, 1994, *Geography: Realms, Regions, and Concepts: 7th Edition.*

PREVIEWING QUESTIONS

1. What does the term doubling time mean?

2. Define perennial irrigation. What kind of structures make this possible?

3. Describe life in the Sultanate of Oman before the development of its oil reserves.

KEY THEMES

- **People and the Land**
- **Gift of the Nile**
- **Egyptian National Planning**
- **Labor in Oman**
- **The Sultan's Vision**
- **Resource Development for the Future**

Egypt: Water shortage + food shortage + over populated

DISCUSSION OF CASE STUDY THEMES

• People and the Land

Sixty million people crowd on 3 percent of the Egypt's total land area. In the video, the soil science professor provides a brief history of the land-to-human ratio of the country. In 1800, there was one acre of land for each person, and, in 1900, it was a half acre per person. Now there is less than one-eighth an acre — less than 600 square yards — per person. This is not enough land to grow enough food to feed the population, a problem that will become ever more complicated if the population doubles as expected in the next quarter century.

• Gift of the Nile

The White Nile's headwaters rise in Uganda and flow through Sudan to Khartoum where it meets the Blue Nile waters from Ethiopia. The longest river in the world, the Nile completes its 4,145-mile journey north to the Mediterranean Sea beyond Cairo where it forms a 115-mile delta. Silt deposited by the Nile's annual overflow has brought agricultural prosperity throughout Egypt's history. Since the early 1960s, the river has been harnessed by the Aswan High Dam to provide constant irrigation.

By the time construction was complete, the main purpose of the dam's regulation was to irrigate more land and to extend the benefits of irrigation to new land. Since then, controlled water supply has allowed for more than one crop per year, and less time between harvests has increased the food supply. Reclaimed desert land has put over a million and half acres into agricultural production.

• Egyptian National Planning

As the desert bloomed, migration to the Nile Valley boomed to more than 5,500 inhabitants per square mile. But, at the same time, cities have grown and taken more than a million acres of prime riverside and delta land out of agricultural production.

Urbanization — the building of centralized services and housing — is one approach to provide for the burgeoning population. New Towns are planned for self-sufficiency. Small industries are enticed through tax incentives to start up near new housing projects. Plans to distribute the population growth are made with an eye to water conservation. A light industry worker is expected to use less water in the city than he did as a fellaheen, or farmer. In this way, concentrating people in urban areas spares the elixir of the desert.

Though water reserves are creating more farmland, the marginal soils of this newly irrigated land have low production. At the same time, the cities built for water resource management plans have actually been built over some high-production fertile soils, unintentionally offsetting the agricultural gain provided by the Aswan Dam's regulation of the Nile.

• Labor in Oman

When oil was discovered in the Arabian peninsula, imported labor was relied on to help extract this export commodity. Expatriates, or guest workers, flooded into Oman in the mid-1960s to fill positions in this new industry in which the domestic population was not trained.

Oil was not only a new industry, but the only industry for several decades. Now that oil money has transformed their society, Omanis are preparing for work in the developing sectors of their modernized economy. Vocational training from office work to welding is the first step to reach the goals set forth by Omanization, a policy to replace 35 percent of the 400,000 guest workers with Omani men and women by the end of 1996.

• The Sultan's Vision

Sultan Qaboos took over a poor and undeveloped country from his father in 1970. He decided to use the money from oil to "bring the people from darkness of the past into a new future." The first immediate change was a nationwide educational program that included women. Education is the cornerstone of the Omanization policy for domestic worker development, which in turn promotes the diversification of the economy.

• Resource Development for the Future

The Sultan's plans for economic diversification are driven by the realization that oil reserves are a finite resource. Oil revenues are being reinvested in fisheries, mineral mining, agriculture and human resources.

A hundred miles southeast of Muscat, the capital city, a modern fishing industry is under development. Fishing has always been the main occupation in this area, but a new road to Muscat and its international airport now provides opportunities for the fishermen to find extended markets. The number of registered fishermen has increased 30 percent over the last few years. Future plans include government investment in new harbors, cold storage and processing facilities. By the end of 1996, it is estimated that the fishing industry in Oman will contribute 10 percent of the country's Gross Domestic Product.

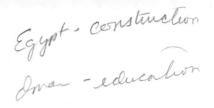

Egypt - construction

Oman - education

CASE STUDY CONNECTIONS

1. Desert conditions dominate the landscapes of both Egypt and Oman. Nearly all of the 3 percent of Egyptian land that is arable lies along the Nile River.

2. Both states depend on their natural resources; Egypt imports 50 percent of its food supply.

3. Nomadic Omanis have recently settled into urban centers, while Egyptian fellaheen inhabit the banks of the Nile.

4. Egypt's development is tied to a fluctuating water source; Oman is planning beyond its finite oil reserves.

5. The focus of long-term national planning is taking the form of construction projects in Egypt and education in Oman.

Egypt - water - lack of

Oman - limited oil

TEST YOUR UNDERSTANDING

Questions

1. How does an increasing population in Cairo affect the prime agricultural land in the Nile Delta?

2. At current rates, how soon is the population of Egypt expected to double?

3. What is the relationship between modernization and the discovery of oil in Oman?

4. Did guest workers affect the culture of Oman? How?

Map Exercises

The maps provided at the beginning of this unit can be used to complete this exercise. You may find extra copies of these maps helpful. Also, refer to the maps of Chapter 7, "North Africa/Southwest Asia: The Challenge of Islam," from de Blij and Muller, 1994, *Geography: Realms, Regions, and Concepts: 7th Edition*, and any standard atlas to assist in this exercise.

1. Place the following names:

Egypt	Cairo	The Nile Delta
The Nile	Sudan	Lake Victoria
The White Nile	Ethiopia	The Blue Nile
Uganda	Muscat	Oman
The African Horn	Bab el Mandeb Strait	Gulf of Aden
Arabian Sea	Gulf of Oman	Hormuz Strait
Arabian Peninsula	Yemen	Saudi Arabia

2. Using a blue marker, trace the flow of the White Nile, the Blue Nile and the combined Nile from their headwaters to the Nile Delta. Shade the arable land of Egypt in green. Use red dots to show the population distribution of Egypt.

3. Locate the border between Oman and Saudi Arabia.

CONSIDER THE DISCIPLINE

EGYPT AS A CULTURE HEARTH

The themes of **cultural geography** are five-fold, including the study of cultural **landscapes,** culture **hearths,** cultural **diffusion,** cultural **ecology,** and culture **regions.** Whereas Jerusalem epitomizes one of the world's extraordinary urban cultural landscapes, Egypt is best understood as one of the world's great cultural hearths.

The origins of ancient Egyptian civilization can be traced back more than 5,000 years. In addition to the great pyramids of Giza and the temples of the Upper Nile, Egypt's lasting **legacy** includes the **domestication** of a variety of cereals, vegetables, fruits, and animals, and **technological and scientific advances** such as irrigation, mathematics, metallurgy, and engineering.

From the perspective of cultural geography, the significance of these achievements is their genesis and diffusion throughout the realm, and indeed, throughout other places in the world, particularly in Europe. It is not too great a generalization to observe that Egyptian achievements represent one of the **pillars of Western civilization,** identifying this cultural hearth as **pivotal** in **world history.** But Egypt's achievement cannot only be measured by the spread and influence of the country's ideas and scientific knowledge. It is important also to understand the magnitude of advancements made within Egyptian civilization. Since ancient times, Egyptians have applied their technology and know-how to solve the basic problems of human existence, for example, the development of **sophisticated land- and water-use practices** to productively manage the Nile River through **irrigation** of its flood-plain soils in a harsh desert environment.

To further develop your understanding of this field of regional geography, refer to the Focus on a Systematic Field in Chapter 7, "North Africa/Southwest Asia: The Challenge of Islam," from de Blij and Muller, 1994, *Geography: Realms, Regions, and Concepts: 7th Edition.*

CRITICAL VIEWING

Based on the Case Studies in Video Program 18: "Population, Food Supply, and Energy Development" from *The Power of Place: World Regional Geography,* and the reading in Chapter 7, "North Africa/Southwest Asia: The Challenge of Islam," from de Blij and Muller, 1994, *Geography: Realms, Regions, and Concepts: 7th Edition,* develop essays answering the following questions:

1. Research the construction and results of the Aswan High Dam. What are the ecological and economic impacts of this engineering on the Nile? Which planned outcomes have been achieved? Which outcomes were unintended? Consult your local library for further research materials.

2. How do plans for Omanization affect Arab and Islamic traditions? Why has modernization in Oman not given rise to the fundamentalism seen in the Case Study *Istanbul: Fundamental Change?* Might this change in the future?

■ UNIT 9

Subsaharan Africa: Realm of Reversals

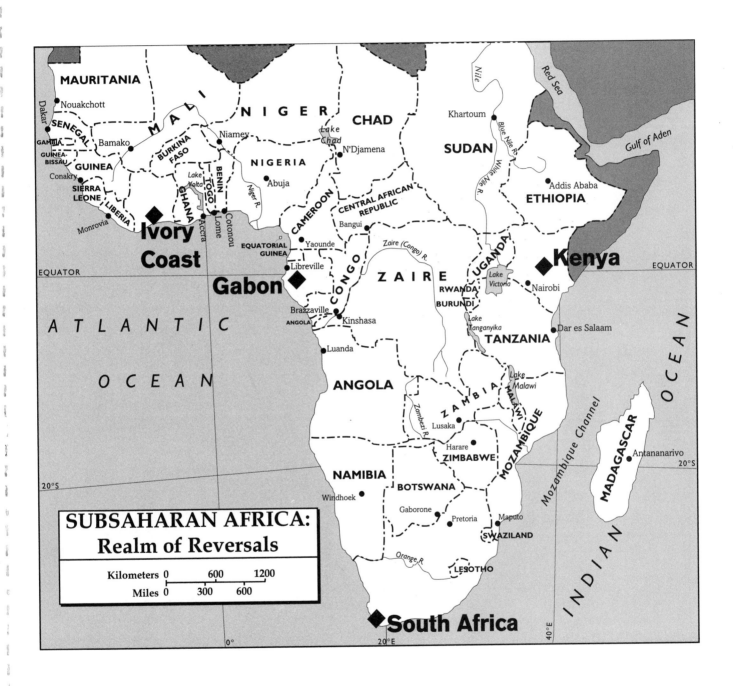

SUBSAHARAN AFRICA:
Realm of Reversals

| Kilometers | 0 | | 600 | | 1200 |
| Miles | 0 | 300 | | 600 | |

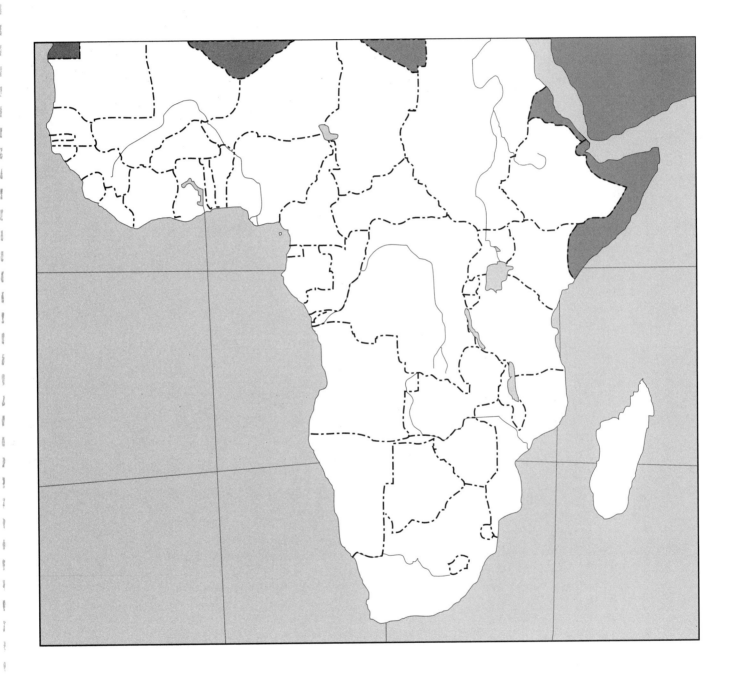

▪ PROGRAM 19

The Legacy of Colonization

LESSONS IN GEOGRAPHY

At the end of Program 19, you should be able to:

1. Discuss the impact of colonialism on present-day Africa.
2. Understand the economic development challenges facing the former French colonies of Ivory Coast and Gabon.
3. Review the physical characteristics of Ivory Coast and Gabon and describe how they affect each country's population distribution.
4. Identify how the international price of commodities affects the living standards for the citizens of both countries.
5. Identify factors which cause migration from rural to urban areas.

OVERVIEW

Program 19 examines *Ivory Coast: The Legacy of Colonialism* and *Gabon: A Future in Oil?* in the realm of **Subsaharan Africa.** Both countries are former colonies of France and have rich, culturally diverse populations. Each country has pursued a path of development with limited success.

Ivory Coast, officially named Cote d' Ivoire, is the largest cocoa producer in the world, and it experienced strong economic growth in the 1960s and 1970s based upon the export of this commodity. **Plantation agriculture** requires a hot, humid rainforest environment for cultivation and demands significant labor for crop production, creating work opportunities for people in and out of the country. **Migration** to fill this demand has resulted in an influx of people from different ethnic groups as seen in the village of Nipy, located in the southwestern part of this West African country.

Agricultural wealth influenced the growth and activities of Abidjan, the principal city and port of Ivory Coast. The government administration centered in Abidjan manages **economic stabilization programs** which have encouraged growth and aided development through investment in the country's **infrastructure.** The creation of services and facilities in the new, more geographically centered capital of Yamoussoukro has had positive and negative aspects, mirroring the situation throughout the country. Major economic, environmental, and social problems have arisen from overdependence on a single commodity, deforestation caused by an increase in cultivation, and an influx of migrant peoples.

The Gabon Case Study looks at another former colonial possession of France. Situated in Equatorial Africa, the country is sparsely populated and its economy is based upon the export of **primary resources**, mostly derived from oil and wood products. While oil exports now dominate in terms of revenue, the exploitation of forested areas to produce wood products remains a concern because 15 percent of the population is employed in this sector.

The Legacy of Colonization   169

must diversify into agriculture not depend just on oil

Like Ivory Coast, the fragile structure of the Gabonese economy is tied to external demand for its commodities. Because falling prices in international markets have decreased domestic tax revenues, government spending at home cannot adequately provide public services and utilities, as seen in the **shanty town** near Libreville, the country's capital and largest city. Most of the residents inhabiting these inadequate quarters and low-lying flood plains have migrated from rural areas. The financial crisis of 1986, caused by the fall of the price of oil and the **devaluation** of Gabonese currency, has been difficult on the poor.

To improve life for the people, the government is attempting to restructure its policy and reduce dependence on oil as the basis for its economy. While oil will remain an important export for the next few years, **diversification** into agriculture and forestry will contribute to the stabilization of the country's economy.

STUDY RESOURCES

1. Video Program 19: "The Legacy of Colonization: Ivory Coast and Gabon," from *The Power of Place: World Regional Geography*.
2. Unit 7 Map, "Subsaharan Africa: Realm of Reversals," from Latz and Gilbert, 1996, *The Power of Place: World Regional Geography*.
3. Chapter 8, "Subsaharan Africa: Realm of Reversals," from de Blij and Muller, 1994, *Geography: Realms, Regions, and Concepts: 7th Edition*.

PREVIEWING QUESTIONS

1. How do the political boundaries of Colonial Africa in 1950 compare to the nation-state boundaries of today?
2. What is the Sahel?
3. Which countries' monetary systems are based on the Communaute Financiere Africaine (CFA) franc?
4. What kind of problems do the physiographies of West and Equatorial Africa present for agriculture?

KEY THEMES

- **The Cacao Plantation System**
- **The Relationship of Agriculture to Urban Growth**
- **The Colonial Imprint of Commodity-Based Economies**
- **Factors Affecting Migration**

DISCUSSION OF CASE STUDY THEMES

• The Cacao Plantation System

The agricultural practices associated with both cocoa and coffee production have shaped the landscape of Ivory Coast over time. Plantations were established in rainforest areas exhibiting the hot, humid climate favored by the cacao plant. During the early development of these plantations, the forest remained intact and sparsely populated, and the plantation's growth depended upon the arrival of migrant workers seeking economic opportunities. Today, one-third of Ivory Coast's population is composed of Burkinabes and other immigrants, originally attracted to work the plantations.

High demographic growth rates of 3 percent per year and the large influx of migrants have expanded the cultivation of cacao further into the forest. Even so, this primary industry is reaching capacity and can no longer provide adequate work for farmers. With the loss of virgin forest land, few opportunities exist for the next generation to support a traditional village lifestyle.

• The Relationship of Agriculture to Urban Growth

The port of Abidjan provides one example of the influence agricultural wealth has had upon the growth and activity of Ivory Coast. This growth is fostered principally by the government's investment in and management of the stabilization fund, a fund which controls the commercialization of cacao and coffee beans, and other agricultural exports. Because minimum market rates are fixed for products across the country, capital is accumulated when the price of commodities is high on the international open market. The accumulated capital is then available for investment, creating industrial growth and strengthening the country's infrastructure. This stimulated economy also creates job opportunities which attract migrant workers. Unfortunately, as these workers move into the city they encounter a service economy which is becoming increasingly overloaded as it attempts to fulfill the needs of a rapidly growing population.

• The Colonial Imprint of Commodity-Based Economies

Because developing countries are dependent upon the export of a narrow range of resources, they are often subject to commodity price fluctuations beyond their control. This difficulty afflicts both Ivory Coast and Gabon because each relies upon the exploitation of natural resources to obtain the funds necessary to improve living conditions within their boundaries. Ivory Coast remains dependent upon cacao bean production, while Gabon is reliant on the export of oil. In an increasingly globalized economy, price fluctuation will become even less predictable, creating an even greater challenge for developing countries.

This pattern of exportation of raw materials by countries with undiversified economies in order to reduce their debt is an extension of colonialism. In the countries featured in this program, ties to the colonial legacy still remain strong, not only in economic terms, but in culture, political processes, and sources of aid. France, which formerly governed both Ivory Coast and Gabon, will play an important role in each country's future standard of development.

• Factors Affecting Migration

As Gabon struggles, people continue to flock to its cities which are unable to provide employment and fulfill basic needs. As these migrant workers arrive, cities lack the urban services (i.e. housing, sanitation, etc.) to support the growing population. This flow of rural-to-urban migration is also evident in Ivory Coast. Despite inadequate infrastructures which are unable to support expanding populations, urban growth continues. This tide of migrants may be stemmed by making improvement to primary production, thereby reducing the high percentage of food products imported to these countries, however, the problems created by these overcrowded conditions cannot be solved simply by adding services and strengthening infrastructure through government investment. There must also be a recognition of the conditions in rural areas which prompt people to choose to migrate. Both Case Studies have identified agriculture and poorly developed service economies as critical factors influencing the migration patterns found in these countries. Additional factors involved in relocation may include political instability and access to both education and healthcare.

CASE STUDY CONNECTIONS

1. Both countries share the CFA franc and the French language as products of their colonial acculturation.
2. Urbanization has developed in both countries around the export of primary resources.
3. Urban migration pressures affect both countries.
4. Economic crises in both nations have been the result of price reductions in world markets, monetary devaluation, and lack of infrastructure.
5. Neither country adds value to the commodities it ships abroad.
6. The limited infrastructure in each country's interior supports the drain of resources.
7. The relative location of each country provides each with opportunities for economic development.

GEOGRAPHERS AT WORK

The relationship between humans and the natural environment is a central focus of geographic research, a field of study which is becoming increasingly important as population pressures, resource exploitation, and human-induced pollution inexorably alter the fragile balance between humans and the earth that sustains them.

Gabonnese geographer Henri Mapaga-Koumba studies human/environment relations. His work on the management of Gabon's forest reserves typifies what is distinctive about the geographic perspective, a holistic view of the environment that incorporates the study of the physical forces that shape the surface of the earth and the variety of ways human beings adapt to and modify the parts of the earth in which they dwell. Such study of the environment is bound up in an effort to understand human interaction with and alteration of the natural environment.

Human alteration of the environment can be documented throughout the developing world of Africa. Reducing the careless exploitation of forest resources is but one of many examples of the resource management challenges facing the country. Others include addressing desertification, strip mining, fishing, hunting, and soil erosion resulting from intensive agricultural activity.

TEST YOUR UNDERSTANDING

Questions

1. What ethnic characteristics do the countries share? Describe the religious affiliations of each country's population.

2. What spatial characteristics do the countries share with respect to population distribution?

3. List three problems created by a commodity-export economy.

4. Describe the link between each country's urban areas and its agricultural sector.

Map Exercises

The maps provided at the beginning of this unit can be used to complete this exercise. You may find extra copies of these maps helpful. Also, refer to the maps of Chapter 8, "*Subsaharan Africa: Realm of Reversals,*" from de Blij and Muller, 1994, *Geography: Realms, Regions, and Concepts: 7th Edition*, and any atlas to assist in this exercise.

1. Place the following names:

Sahel	Niger	Burkina Faso
Libreville	Congo	Nigeria
Senegal	Togo	Cameroon
Zaire	Benin	Central African Republic
Yamoussoukro	Brazzaville	Equatorial Guinea
Ouagadougou	Abidjan	Nipy

2. Based on your reading and maps of Africa's physiography, languages, political units, and colonization patterns, define and map the African Transition Zone.

CONSIDER THE DISCIPLINE

THE AFRICAN AGRICULTURAL HERITAGE

The great majority of Africans live today as their predecessors lived, by **subsistence farming,** herding, or both. The subsistence form of livelihood was changed only very indirectly by **colonialism**; tens of thousands of villages across Africa were never fully brought into the economic orbit of the European invaders, and life in these settlements went on more or less unchanged. Thus, the history as well as the **location** and spatial extent of European colonial activity must be factors in any consideration of Africa's agricultural prospects.

After more than four centuries of contact, Europe finally laid claim to all of Africa during the second half of the 19th century. Geographers are interested in the ways in which the philosophies and policies of colonial powers were reflected in the **spatial organization** and **political infrastructure** of African dependencies and the more contemporary independent states. An example of this is the European presence on the West African coast which brought about a complete reorientation of trade routes, initiated the decline of the interior savanna states, and strengthened the coastal forest states. Within this region, France's colonial influence is instructive. France, the colonial power whose sphere of influence included Ivory Coast and Gabon, is the classic example of the **centralized unitary state**. As noted in the de Blij text, the French pattern in which all roads lead to Paris was repeated in colonial Africa, where all roads led to France, French culture, French institutions.

A gradual change in the involvement of West Africans in a **cash economy** has taken place during the past three decades. Excluded during the colonial period from most activities that could generate income for themselves and their families, many Africans have chosen, since independence, to make changes that promise to bring in money. Farmers have introduced cash crops onto their small plots, or, in the case of Ivory Coast, exploited extensive groves of the cocoa-producing cacao trees; at times the search for cash for farming has replaced subsistence crops altogether. Even so, Africa continues to fall farther behind in meeting its food needs, and, even in areas like Ivory Coast where the promise of cash crop farming is brightest, when international prices decline, so too does the ability of the farmer, and indeed the national government, to support and purchase, much less to grow essential, life-sustaining foods.

For a detailed discussion of African agriculture and colonial spatial organization, refer to Chapter 8, "Subsaharan Africa: Realm of Reversals," from de Blij and Muller, 1994, *Geography: Realms, Regions, and Concepts: 7th Edition*.

CRITICAL VIEWING

Based on the Case Studies in Video Program 19: "The Legacy of Colonization," from *The Power of Place: World Regional Geography*, and the reading in Chapter 8, "Subsaharan Africa: Realm of Reversals," from de Blij and Muller, 1994, *Geography: Realms, Regions, and Concepts: 7th Edition*, develop essays answering the following questions:

1. Describe the effects of colonialism on each country and provide examples of its legacy in modern times.

2. Can the pressures created by urbanization and population growth be eased through increased agricultural productivity as suggested by the de Blij text's box "A Green Revolution for Africa?"

3. Nationalism, devotion to one's country, group, or "nation," is a phenomenon found in Africa as well as in other parts of the world. Describe the conditions in Gabon and Ivory Cost which could influence the development of nationalism.

4. Add climate and vegetation patterns to the map of the African Transition Zone created in Map Exercise 2, then analyze the possible regional complementarity in both countries. What role would infrastructure play in the exchange opportunities in each area?

■ PROGRAM 20
Understanding Sickness, Overcoming Prejudice

LESSONS IN GEOGRAPHY

At the end of Program 20, you should be able to:

1. Discuss medical geography and the delivery of healthcare in Kenya.
2. Identify the difficulties involved in providing public healthcare in a developing country.
3. Explain the South African plan to reduce poverty through land reform.
4. Understand the potential conflicts which can develop during the implementation of new government policy.
5. Forecast the continental leadership potential for the new South Africa.

OVERVIEW

The *Kenya: Understanding Sickness* Case Study examines **medical geography** and the delivery of healthcare in a **developing country**. The central question asked in this program is: is it possible to improve the situation for people in Kenya, where conditions seem to be more favorable for the development of disease than for the survival of human beings? The program highlights three diseases afflicting Kenya and the Subsaharan region of Africa: **Acquired Immune Deficiency Syndrome (AIDS),** malaria, and trachoma.

South Africa: This Land is My Land is a complex study which began to unfold after the country's first democratically elected government, headed by President Nelson Mandela, came to power in 1994. The new government faces many challenges in developing **post–apartheid policy**, including reducing poverty, promoting community development, and fostering political stability. This Case Study addresses the reform of **land policy** remaining from apartheid in an effort to improve the lives of the greater population.

Both South Africa and Kenya face significant challenges in their struggle to improve the lives of their respective populations. This program reviews two separate areas, each of which have specific development problems. In Kenya's case, **environmental conditions** favor a wide range of diseases. Efforts to control these environmental factors are complicated by many conditions common to developing countries: poor transportation and communication networks, a low education level, and a limited availability of basic community services.

For South Africa, the challenge is to alleviate poverty by reversing the legacy of apartheid, a government racial policy that persisted for decades. The implementation of **land reform** as a method to reverse the effects of concentrated

lower life expectancy

ownership by whites is an attempt to alter problems created not by environmental conditions but by societal structures. To achieve equity and diversity, the new government is pursing a legal framework to **redistribute** land ownership. A process creating more individual farms is likely to affect the landscape due to smaller plot size, crop diversity, and more labor-intensive methods of agriculture. The threats and tensions which arise during the implementation of land reform create a basis for potential conflicts between not only blacks and whites, but among a wide range of **multi-ethnic groups** as the government works to find a fair method of redistributing land to redress past injustices and reduce the level of contemporary poverty.

STUDY RESOURCES

1. Video Program 20: "Understanding Sickness, Overcoming Prejudice," from *The Power of Place: World Regional Geography.*
2. Unit 9 Map, "Subsaharan Africa: Realm of Reversals," from Latz and Gilbert, 1996, *The Power of Place: World Regional Geography.*
3. Chapter 8, "Subsaharan Africa: Realm of Reversals," from de Blij and Muller, 1994, *Geography: Realms, Regions, and Concepts: 7th Edition.*

PREVIEWING QUESTIONS

1. What was the policy of apartheid?
2. How do the occurrence and transmission of disease differ among developed and developing countries?
3. What problems do developing countries typically encounter when creating a healthcare system for a geographically dispersed and largely rural population?

KEY THEMES

*   **Medical Geography**
*   **The Delivery of Healthcare**
*   **The Relationship of South Africans to the Land**
*   **The Alleviation of Poverty through Land Reform**
*   **Potential Changes in the Agricultural Sector**

DISCUSSION OF CASE STUDY THEMES

• Medical Geography

The diseases afflicting the population of Kenya are found in many parts of Africa. Environmental conditions in Subsaharan Africa favor a wide range of infectious diseases, creating a deadly threat and causing suffering in the human population. People living in the rural, isolated, and undeveloped areas of Kenya are particularly susceptible to contracting diseases. Because these areas lack adequate services, infrastructure, and on-site healthcare, efforts to control the transmission of disease are difficult. Human factors, such as poor hygiene, further aid the spread of disease, making some diseases more common. To change these cultural patterns, human behavior must be modified.

• The Delivery of Healthcare

Strategies to address public health rest with the government of Kenya and present enormous challenges for the healthcare community. Diseases like malaria and trachoma are still prevalent in the rural areas which house nearly 80 percent of Kenya's population, and the expanding number of people afflicted by AIDS and infected with the HIV virus is likely to place a further strain on healthcare programs. To reach these rural locations, healthcare workers must travel to remote areas supported by poor transportation services. They rely on different approaches to deliver care to the people and depend upon educational methods to change patterns of behavior which spread disease. By addressing healthcare needs at a community level, the Ministry of Public Health is attempting a broad approach to improve the lives of Kenya's people.

• The Relationship of South Africans to the Land

The forced migration of black Africans from the lands of their birth is deeply embedded in South Africa's turbulent history. Dictated by apartheid, this movement of people to undeveloped rural lands severed black Africans' connection to their homelands. These people, displaced by a policy created by the white-only government, have long endured unemployment, poverty, and repression. Because the black majority is forced to earn its living from just 15 percent of the land, a great disparity of wealth exists between whites and blacks. After they have withstood generations of suffering as a result of forced migration, it is difficult to compensate black Africans through a normal civil process. The displaced people still feel a strong bond to their places of origin, and returning to their homelands is a powerful, emotional experience.

• The Alleviation of Poverty through Land Reform

The Land Reform Program proposed by the new South African government is designed to address the high level of rural poverty and reverse the effects of apartheid policies. This process is not easy given the threat of great financial loss to the wealthy, most of whom are white; and many issues arise regarding the fairness of land redistribution. The process is further complicated by the multi-ethnic composition of the population and the complex claims each group makes regarding ownership.

As they prepare to redistribute the land among the black majority, government officials emphasize the importance of fairness. They must be sure that people given access to South Africa's resources will use the land to benefit the country as a whole. Land distribution near urban areas is likely to be especially difficult. These densely settled suburbs have been traditionally inhabited and owned by whites; blacks were legally only tenants and had no ownership rights. Because blacks lacked this right to private ownership, whites do not readily accept their claims to the land.

• Potential Changes in the Agricultural Sector

Policies reversing recent historic increases in farm size are likely to be created because smaller holdings provide greater options for a more intensive and diversified agricultural system. These efforts to create a sustainable agricultural system are impaired by a lack of suitable farmland and the resistance of white farmers to give up their farms.

The problems described in interviews with government officials outline firm positions which have been molded by government policies which affect the spatial characteristics of land ownership. The white farmer's fear of losing land is very real, and, due to the size of the farms, his potential loss is great. Between 1950 and 1980, farm sizes almost doubled as a result of government policy. The smaller-scale agriculture planned in the Land Reform Program is based upon more intensive and diversified systems and will be a policy which departs from the older government's position.

CASE STUDY CONNECTIONS

1. In both Kenya and South Africa, rural poverty affects the health and welfare of a large percentage of the population.

2. In each country, government is expected to play a large role in the resolution of social and economic problems associated with sickness and prejudice.

3. Throughout Subsaharan Africa, agriculture is the dominant mode of economic activity in terms of numbers of laborers. In South Africa, however, the economy is more diversified and robust; development opportunities exist in sectors other than agriculture.

4. Ethnic diversity is a common denominator in both countries, with definable ethnic groups concentrated in particular places.

5. Education of the population is a key challenge facing each country.

6. Resistance to change, as seen in Kenya, and preservation of the status quo in South Africa challenge the maintenance of social order.

TEST YOUR UNDERSTANDING

Questions

1. What historic factors have influenced land ownership patterns in South Africa?

2. Why is land reform important to South Africa's political and economic future?

3. List five diseases afflicting Kenya's population.

4. Explain the different physical geographic characteristics that distinguish South Africa from Kenya.

5. List three problems hindering the progress of healthcare for Kenya.

Map Exercises

The maps provided at the beginning of this unit can be used to complete this exercise. You may find extra copies of these maps helpful. Also, refer to the maps of Chapter 8, "Subsaharan Africa: Realm of Reversals," from de Blij and Muller, 1994, *Geography: Realms, Regions, and Concepts: 7th Edition*, and any standard atlas to assist in this exercise.

1. Place the following names:

South Africa	Kenya	Gabon
Pretoria	Nairobi	Libreville
Johannesburg	Lake Victoria	Ivory Coast
Orange River	Mombasa	Abidjan
Vaal River	Lake Turkana	Gulf of Guinea
Cape of Good Hope	Mt. Kenya	Sahel Zone
Indian Ocean	Atlantic Ocean	

CONSIDER THE DISCIPLINE

MEDICAL GEOGRAPHY

The World Health Organization estimated in 1994 that about 18 million people worldwide were infected by HIV; 11 million of them lived in Subsaharan Africa. Kenya's population has been severely affected by HIV, and many deaths have resulted from **AIDS**. In rural areas, almost half of the population between the ages of 15 and 40 has been infected by HIV or has died from AIDS-related complications.

To address the AIDS crisis in Kenya, health organizations try to reach people not only with treatment, but through education. Members of high risk groups include truck drivers and other mobile workers and their sexual partners. Hotel keepers and owners are deputized by health workers to help educate and influence members of these groups in order to modify their high-risk behavior.

The review of trachoma and malaria in Program 20 illustrates the relationship between **environmental factors and humans**. Trachoma is a communicable disease which causes blindness and a great deal of suffering in rural villages. The disease favors dry, dusty conditions and flourishes in these villages due an abundance of fecal matter from animals, such as donkeys, as well as flies and poor human hygiene. Most of those infected are women and children. Simple preventative techniques, such as washing the face, are difficult since water must be obtained and carried by a donkey over a distance of nearly 10 miles (16 kilometers). Prevention of the disease hinges on changing the behavior of people living in the villages and improving social services like water delivery systems. Health workers realize that they must involve themselves in the community in order to implement changes.

Malaria, the most common **vector-borne disease**, afflicts more than 300 million people, 90 percent of whom live in tropical Africa. In Subsaharan Africa the most severe type of malaria, *falsiforum malaria*, is hosted by a specific parasite and transmitted by a species of mosquito, the vector, which can inflict the disease with a single bite. The spread of this disease is exacerbated by a human population weakened by a lack of proper nutrition and adequate services. Health services are not yet very effective, as they remain poorly developed due to a lack of infrastructure and funds.

Prevention rests with understanding the combination of factors influencing the spread of the disease. Controlling the breeding of mosquitoes where people live is an effective and cheap way to alter the environmental conditions which favor the insect. The effort by the village shown in the video to eradicate breeding grounds provides an example of this technique. Governmental health organizations have also started to provide people with bednets which are locally manufactured and distributed. Once again, this process involves changing human behavior. It also raises questions regarding how these new methods may affect the **natural immunity** of the population.

To further develop your understanding of this field of regional geography, refer to the Focus on a Systematic Field in Chapter 8, "Subsaharan Africa: Realm of Reversals," from de Blij and Muller, 1994, *Geography: Realms, Regions, and Concepts: 7th Edition*.

CRITICAL VIEWING

Based on the Case Studies in Video Program 20 "Understanding Sickness, Overcoming Prejudice," from *The Power of Place: World Regional Geography,* and the reading in Chapter 8, "Subsaharan Africa: Realm of Reversals," from de Blij and Muller, 1994, *Geography: Realms, Regions, and Concepts: 7th Edition*, develop essays answering the following questions:

1. What historic events influenced the placement of the current boundaries of Kenya, South Africa and many other African nations.

2. Some political geographers believe that future ethnic conflicts are more likely to occur as a result of multi-ethnic divisions in South Africa rather than simply black vs. white. What are reasons to support this perspective?

3. Medical geography is concerned with two major areas: disease ecology and the delivery of healthcare. Using Kenya as an example, expand upon these two terms.

▪ UNIT 10

South Asia: Aspiring India

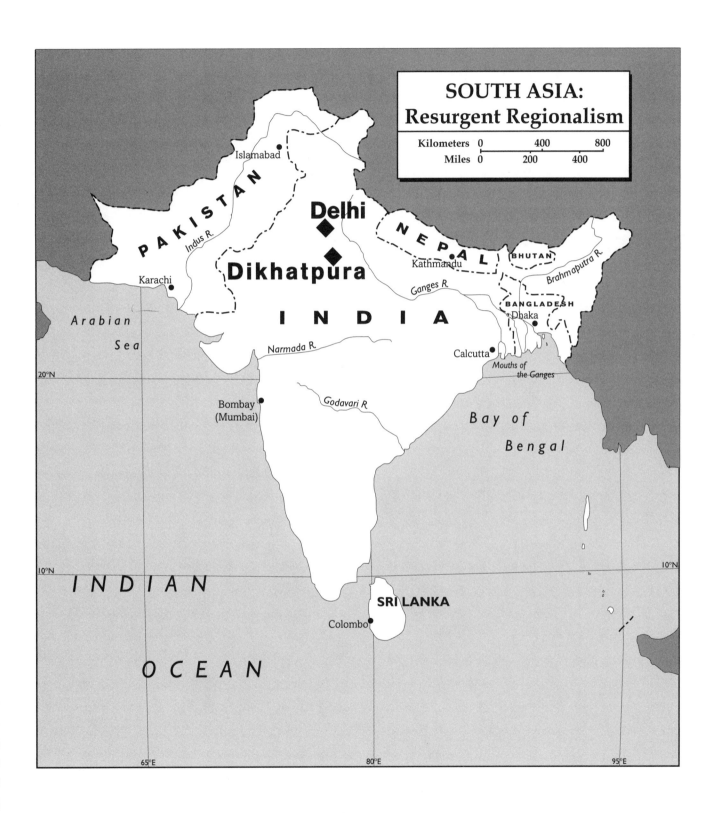

SOUTH ASIA:
Resurgent Regionalism

Kilometers 0   400   800
Miles 0   200   400

PAKISTAN

Islamabad

Karachi

Indus R.

Delhi

Dikhatpura

INDIA

*Arabian
Sea*

20°N

Bombay
(Mumbai)

Narmada R.

Godavari R.

NEPAL

Kathmandu

Ganges R.

BHUTAN

Brahmaputra R.

BANGLADESH

Dhaka

Calcutta

*Mouths of
the Ganges*

*Bay of
Bengal*

INDIAN

10°N   10°N

SRI LANKA

Colombo

OCEAN

65°E   80°E   95°E

▪ PROGRAM 21
Urban and Rural Contrasts

LESSONS IN GEOGRAPHY

At the end of Program 21, you should be able to:

1. Discuss the factors which have fueled Delhi's economic growth.
2. Identify the factors influencing in-migration to the metropolis of Delhi from rural areas.
3. Understand the pressures of population growth on India's capital city.
4. Review the efforts underway that are attempting to plan the regional territory surrounding Delhi.
5. Identify the climatic patterns affecting agriculture in central India.
6. Understand the steps being taken by the Indian government to alleviate rural poverty.
7. Describe the importance of water resource management and its role in reducing the push and pull factors of urban migration.

OVERVIEW

Unit 10 "South Asia: Aspiring India" presents two Case Studies representing **urban and rural** settings in modern India.

Program 21: "Urban and Rural Contrasts" follows a geographer studying the biggest challenge facing one of the world's most populous countries — mass migration to the city. In *Delhi: Bursting at the Seams* potential solutions for the **pull factors** facing a city experiencing **rapid growth** are presented. Squatters camps are now home to more than one-third of Delhi's 9.8 million people, and continued migration threatens to overwhelm the inadequate infrastructure of the city. Several potential solutions are presented as the region's **planning** board struggles to accommodate growth through the design of satellite cities and the attraction of foreign direct investment ventures.

Rural India is presented in *Dikhatpura: Help through Irrigation,* an examination of a village in the drought-prone area of the central Indian state of Madhya Pradesh. The geographer in the video interviews local farmers, who are dependent upon irrigation canals for their livelihood. The case reveals rural India's economic transformation resulting from stabilization of the **agricultural water supply** through construction of an irrigation infrastructure. Despite evidence that the introduction of irrigation has significantly **reduced poverty**, many resulting water **resource management** issues have yet to be resolved. Problems include frequent disputes over the allocation of water among farmers, leakage into the existing water table creating boggy soil, and salinization, the build-up of salt in the soil, which over time renders the soil useless for farming.

STUDY RESOURCES

1. Video Program 21: "Urban and Rural Contrasts" from *The Power of Place: World Regional Geography*.

2. Unit 10 Map, "South Asia: Resurgent Regionalism," from Latz and Gilbert, 1996, *The Power of Place: World Regional Geography*.

3. Chapter 9, "South Asia: Resurgent Regionalism," from de Blij and Muller, 1994, *Geography: Realms, Regions, and Concepts: 7th Edition*.

PREVIEWING QUESTIONS

1. Why would a national government assume responsibility for providing sophisticated irrigation networks to selected agricultural areas?

2. What factors lead farmers to leave villages for urban areas?

3. What are the problems associated with rapid urban growth and how do regional governments attempt to resolve these probelms?

4. What are the defining characteristics of the new middle class in a developing country?

KEY THEMES

- **Rural-to-Urban Migration**
- **The Problems of Metropolitan Delhi**
- **Regional Planning Efforts**
- **Effects of the Subcontinent's Climate on Indian Agriculture**
- **Water Management and Its Role in the Development Process**
- **Effects of Irrigation Development**
- **Improving Life in Villages Eases Problems in the City**

DISCUSSION OF CASE STUDY THEMES

• Rural-to-Urban Migration

Delhi, India's capital city, has been undergoing rapid change driven by strong economic growth. Associated with this growth are greater access to goods and services, new cultural activities, and more employment opportunities. The city pulls workers from the countryside like a magnet, but jobs are not available for all the migrants searching for work, and, because housing this expanding group is a significant problem, many people are forced to live in squatter's camps. Most new residents have come from agricultural villages in search of a cash income. They fill many unskilled jobs and typically send a portion of their income back to relatives at home.

• The Problems of Metropolitan Delhi

The changes prompted by in-migration of the population from rural areas has created a severe lack of affordable housing and has contributed to growing squatter communities which now house over one-third of Delhi's residents. Continual population pressures have resulted in shortages of water, electricity, and sewage systems. The number of cars has also exploded, and air pollution and traffic congestion are major problems. While Delhi's economy continues to expand, the sheer volume of in-migrants threatens the city's ability to cope in the future.

• Regional Planning Efforts

The National Capital Region Planning Board is attempting to solve the problems facing Delhi. In a comprehensive plan formulated for the entire capital region, the current size of the metropolitan area is expected to expand by 20 times the existing area. Within the new capital region several satellite cities, connected by expanded rail and highway systems, are planned. The city of Gurgaon, 18 miles (30 kilometers) south of Delhi, is a crucial part of this plan. The city has many factories and housing complexes, and employment there allows more workers to enter a growing middle class. The National Capital Region Plan extends beyond Gurgaon, and the construction of buildings and roads over the entire region continues to both accomodate and attract laborers in search of a cash income.

• Effects of the Subcontinent's Climate on Indian Agriculture

In central India, life in farming villages of Madhya Pradesh revolves around the arrival of the wet monsoons. Normally, rains will fall for four months beginning in October and then are followed by a long dry season. During the dry season, which lasts six to eight months, little rain falls. Because of the scarce precipitation, farmers must rely on stored water or intermittent rains. Droughts are common in India, and over 33 percent of the land mass is subject to only a limited rainfall.

• Water Management and Rural Development

Allaying the fear of drought and reducing poverty have been concerns of the Indian government since its creation in 1947. Modern river development projects based upon building dams for water storage and canals for irrigation were initiated in the late 1950s and 60s. The State of Madhya Pradesh was one of the first to benefit from these national programs. Dikhatpura was first provided with water for irrigation in 1971 and has markedly increased agricultural productivity since then. The additional water supports two harvests per year, and a variety of crops can now be grown. Many farmers were able to move away from subsistence crops into cash crops after following the government's agricultural plan, called the "Green Revolution." Increased productivity and land reform have raised the income of much of India's rural population, extremely poor prior to these changes. With added income and support from the government, many farmers can also raise water buffalo which provide milk which can be sold in nearby towns. This process not only increases the income of rural families, but creates new jobs and a new food source to further alleviate poverty. The changes prompted by the addition of milk as a source of food and income in rural India has been called the country's "White Revolution."

• Effects of Irrigation Development

Modern water resource management techniques have created some unforeseen difficulties in rural India. While government investment created the irrigation network, local control is not always clearly administered. Disputes, illustrated in the video by the altercation between the young boy and the elder farmer, often erupt over the allocation of water. As stated earlier, unlined canals allow leakage into the existing water table which in turn creates boggy soil. Overuse of irrigated water can lead to salinization, the build-up of salt in the soil, which over time renders the soil useless for farming.

• Improving Life in Villages Eases Problems in the City

The development of a stable water supply and delivery system has been an important factor in reducing the poverty found in the rural areas of India. To continue to reduce the push factors driving migrants to the city, further improvements raising the standard of living in these areas will have to be made. The agricultural sector remains the dominant employer of the rural population, and this sector's productivity is tied to water management. By raising income levels and providing services to rural areas, incentives to move to urban areas are reduced. Water management in rural areas can be seen as a key to helping solve the enormous problems facing cities like Delhi, and may even eliminate critical problems like population growth and food shortages in the future.

CASE STUDY CONNECTIONS

1. Rapid population growth is a problem in both urban and rural areas of India.
2. Careful water management in rural India will affect the growth management opportunities in urban India.
3. India's rural and urban communities display both increasing numbers of poor and increasing numbers of residents with disposable income.
4. The role of government planners in urban and rural India is large and of increasing importance.
5. Conflicts over resource management, e.g., air, land, and water, threaten to stifle the growth of the Indian economy.
6. As village life improves, urban problems will decrease.

GEOGRAPHERS AT WORK

Minamino Takeshi, a geographer from Hiroshima University in Japan, tackles in his research one of the most vexing of India's problems, the management of population growth and food supply.

Every country is comprised of networks of economic trade interdependence between urban and rural areas. No one urban center has all the resources it needs to survive. The markets for agricultural products that can be found in most cities represent one example of such mutually beneficial relationships.

Minamino notes, however, that in a developing country like India, before seeking trade opportunities in cities, agricultural producers must first meet local subsistence needs. A prerequisite for this is that local production sites must have in place the supporting infrastructure to create food surpluses. As an economic activity, agriculture depends on capital, resources, power supplies, labor, information and land. In India's case, water for irrigation is an essential part of the production equation, as is maintenance of some kind of balance between population growth and the capacity of a local area to feed itself. From a geographic perspective, it is impossible to imagine the prospects for India's cities and farms independent of one another.

TEST YOUR UNDERSTANDING

Questions

1. List two outside cultures which have left an historic imprint upon Delhi's landscape.
2. What are the factors that have contributed to Delhi's economic growth?
3. Provide a brief explanation of the terms Green Revolution and White Revolution.
4. List three problems associated with managing water in the rural areas of India.

Map Exercises

The maps provided at the beginning of this unit can be used to complete this exercise. You may find extra copies of these maps helpful. Also, refer to the maps of Chapter 9, "South Asia: Resurgent Regionalism," from de Blij and Muller, 1994, *Geography: Realms, Regions, and Concepts: 7th Edition*, and any standard atlas to assist in this exercise.

1. Place the following names:

Dikhatpura	Madhya Pradesh	Chambal River
Ganges River	Vindya Range	Yamuna River
Konkan Coast	Malabar Coast	Bay of Bengal
Western Ghats	Arabian Sea	Territory of Delhi
Central Indian Plateau	Delhi	Gurgaon

2. Diagram the wind flow pattern of the wet monsoon affecting the Indian subcontinent and briefly describe this process. Be sure to reference the major physical geographic features influencing monsoonal movement.

CONSIDER THE DISCIPLINE

POPULATION GEOGRAPHY

India is concerned with its **population** and **food supply problems.** India is the seventh largest country in area and the second most populous country in the world. With **demographers** expecting India's population to exceed that of China by the early 21st century, the successful resolution of these problems is of interest to the world community. Resolution does not rest exclusively with urban growth and rural development, but with a variety of issues.

With a population of over 10 million, Delhi is made up of many different ethnic, religious, and linguistic groups. The city's physical location near two large rivers, the **Ganges** and the **Yamuna**, has attracted people for thousands of years, and the influence of various cultures upon the city is evident. The video discusses the mark left by the **Islamic culture** in the 11th century on the old section of the city, Old Delhi, and on New Delhi, built during the **British Colonial period** at the end of the 19th century.

Since its independence from Britain in 1947, India has become the world's **largest parliamentary democracy** and has achieved considerable social and economic progress based on **agricultural improvements, land reform,** and large-scale **development projects.** Building upon the vast infrastructure developed by the British, during the colonial period, **industrial development** has grown and now provides 25 percent of the country's **Gross Domestic Product.** An expanding **middle class** is altering the face of Indian society even though a proportion of the population still live below the **poverty line.**

To further develop your understanding of this field of regional geography, refer to the Focus on a Systematic Field in Chapter 9, "South Asia: Resurgent Regionalism," from de Blij and Muller, 1994, Geography: Realms, Regions, and Concepts: 7th Edition.

CRITICAL VIEWING

Based on the Case Studies in Video Program 21: "Urban and Rural Contrasts," from *The Power of Place: World Regional Geography,* and the reading in Chapter 9, "South Asia: Resurgent Regionalism," from de Blij and Muller, 1994, *Geography: Realms, Regions, and Concepts: 7th Edition,* develop essays answering the following questions:

1. Describe the environmental threats which are faced by development projects in rural India.

2. What are the effects, both positive and negative, of the expanding middle class on Indian society?

3. India can be described as one of the world's largest agrarian countries or as one of the world's emerging industrial nations. Choose one and provide evidence to support your position.

■ UNIT 11
China and Its Sphere

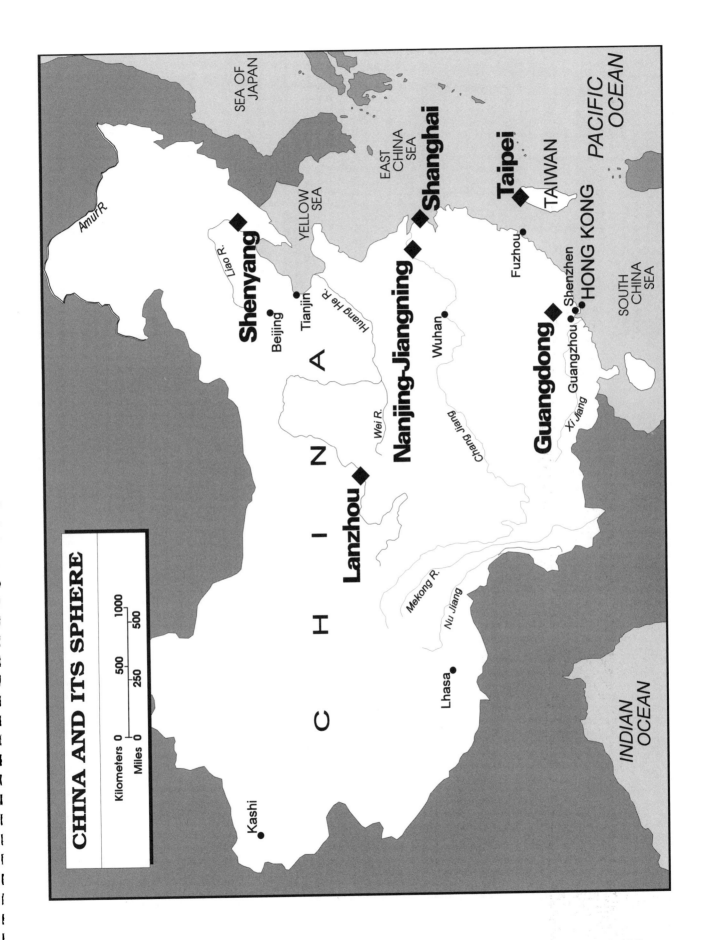

CHINA AND ITS SPHERE

Kilometers 0   500   1000
Miles   0 250 500

SEA OF JAPAN

Amur R.

Shenyang ◆

Liao R.

Beijing ●

Tianjin ●

Huang He R.

YELLOW SEA

EAST CHINA SEA

Shanghai ◆

Nanjing-Jiangning ◆

Wei R.

Wuhan ●

Chang Jiang

Fuzhou ●

Taipei ◆

TAIWAN

PACIFIC OCEAN

Shenzhen ●

HONG KONG

Guangdong ◆

Guangzhou ●

Xi Jiang

SOUTH CHINA SEA

C H I N A

Lanzhou ◆

Mekong R.

Nu Jiang

Lhasa ●

Kashi ●

INDIAN OCEAN

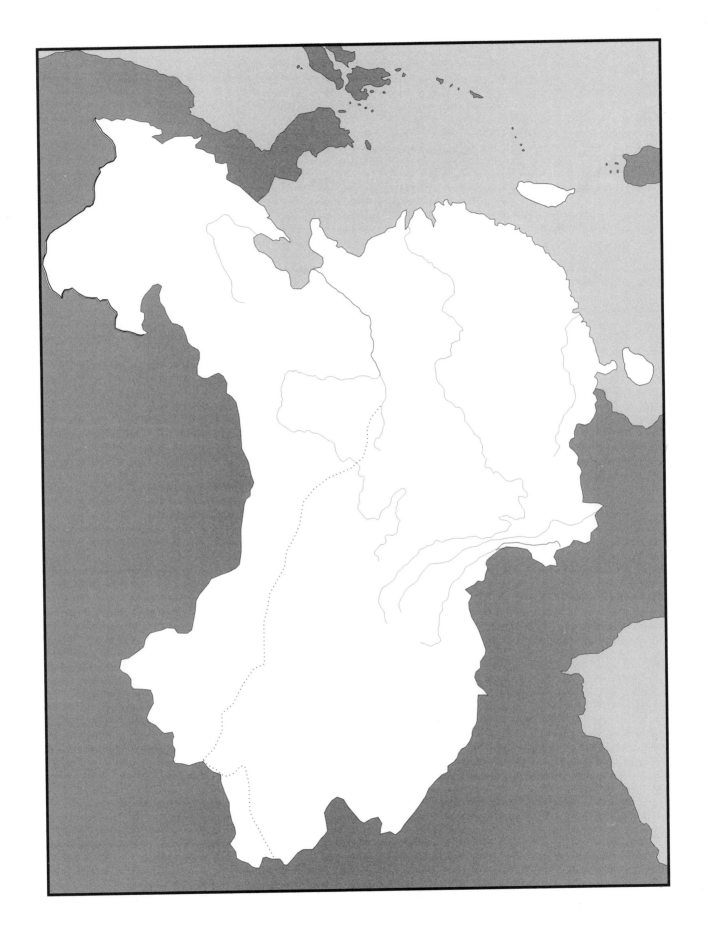

■ PROGRAM 22

Life in China's Frontier Cities

LESSONS IN GEOGRAPHY

At the end of Program 22, you should be able to:

1. Identify the geographic features which influenced the location of Lanzhou and Shenyang.
2. Review the environmental factors that have influenced Lanzhou's development and farming practices.
3. Identify the relationship between Shenyang's natural resources and the location of its industries.
4. Understand a centrally planned economy and some of the factors that are involved in its planning.
5. Examine Shenyang's prospects for future economic growth.

OVERVIEW

In this program two cities on the periphery of China, Lanzhou and Shenyang, are explored. Both are removed from the heartland of Han China by distance and a variety of **geographical barriers.** For Lanzhou, the steppes, plateaus, and mountains of central and western China are barriers, while for Shenyang it has been the Yellow Sea and the broken edge of the Mongolian Plateau. Scenes in *Lanzhou: Confluence of Cultures* illustrate the physical geography of the region, the limits placed on agriculture due to climate and soil, the historic roots of the mixing of different cultures and their influence on modern society. On the other hand, *Shenyang: Hope for China's Rust Belt?*, explores the process of **industrialization** and its decline and revitalization by examining the roles of colonialism, **central planning** and relative location factors. The two cities are in regions of China that have been repeatedly overrun by foreign invaders, and these temporary rulers have left their traces in the **cultural mosaic** of the population and on the **landscape.** The diversity left behind is unique in China, a country where tradition and family ties are deep and have led to the largely homogeneous Han culture.

STUDY RESOURCES

1. Video Program 22: "Life in China's Frontier Cities" from *The Power of Place: World Regional Geography*.
2. Unit 11 Map, "China and Its Sphere" from Latz and Gilbert, 1996, *The Power of Place: World Regional Geography Study Guide*.
3. Chapter 10, "China: The Last Empire?" from de Blij and Muller, 1994, *Geography: Realms, Regions, and Concepts: 7th Edition*.

PREVIEWING QUESTIONS

1. What physical and human factors have contributed to defining the modern city of Lanzhou?

2. How has history shaped Lanzhou's ethnic composition, economy and role in western China?

3. What role has foreign investment played in the development of Shenyang's natural resources and the building of its infrastructure?

4. How has central planning affected Shenyang's growth and decline?

KEY THEMES

- **Lanzhou's Location and Cultural Composition**
- **Economic Growth**
- **Farming Practices**
- **Urbanization**
- **Shenyang's Industrial Development**
- **The Future for Shenyang**

DISCUSSION OF CASE STUDY THEMES

• Lanzhou's Location and Cultural Composition

Lanzhou has been a traditional regional center for thousands of years, serving as a gateway to Central Asia by way of the old Silk Road. The city's location takes advantage of the shallowness of the Huang He as the river leaves the mountains and enters the Lanzhou basin. The confluence of Han, Hui, Mongol, and Tibetan cultures begun in ancient times continues today as the region begins industrial development. The region represents a borderland or frontier which has been under the control of different groups over time. The cultural mosaic found today reflects this frontier situation.

• Farming Practices

Traditional labor-intensive agricultural methods are practiced in the dry, mountainous area surrounding Lanzhou. The unique "stone field" practice seen in the film provides an example of an ancient creative method for preventing the fine loess soil from drying out and blowing away. Adaptation to the semi-arid environment, soil conditions, and steep topography formed the basis of agriculture supporting the region's traditional life style.

• Economic Growth

Modern industrial development began in the early 1950s with the arrival of rail to Lanzhou, an early decision of the Communist government. Industrial development was based upon exploitation of energy resources, notably petroleum. The city now contains large modern industrial plants producing petrochemicals, drilling and refining equipment, as well as chemical fertilizers. Modern agricultural techniques, i.e., pump irrigation, coupled with traditional methods produces a wide range of crops including wheat, barley, millet, corn, soy beans and a variety of fruits supporting local demand.

• Urbanization

Industrialization and agricultural production have supported the city as it grows into an urbanized metropolis. Serving as an urban center for the developing region, the city maintains its links to the past. Use of natural resources, modernization of agricultural practices, and a newly developed energy-intensive industrial economy support the growth of this city isolated in a harsh environment.

• Shenyang's Industrial Development

Covering an area of some 25.5 square miles (66 square kilometers), one third of the land within the city, the Tiexi industrial park has the highest concentration of heavy industry in all of China. Chief among the 950 factories in the park are machinery manufacturing and chemical processing plants. The Liaoning Province in Northeast China is blessed with large deposits of raw materials such as coal and iron. These natural resources, in part, led Japan to invade China in the 1930s in order to fuel its own economy. While occupying the region, the Japanese invested in the development of much of the industrial infrastructure, such as rail and road networks, that exists today.

• The Decline of Shenyang's Economic Might

Economic growth in Shenyang lags behind the rest of China, particularly the booming maritime Southeast. When tensions between China and the Soviet Union were high in the 1950s and 60s, investment was diverted out of the region and into the south. Communist agrarian reform also caused a decline in investment in the region's industrial base. In recent years, many of the facilities have become outdated, and little new construction has occurred to reverse this trend.

• The Future for Shenyang

With the disappearance of many of the traditional machine shops which provide equipment for heavy industry throughout Japan, Korea, and most of China, Shenyang will face stiff competition even as it takes advantage of its skilled workforce and sturdy infrastructure. Recently, Volkswagen, the German auto manufacturer, began production at Jilin Province's massive Automobile Factory #1. The factory is reported to employ or have on pension over 200,000 workers. A bright spot in Shenyang's economic future is its highly skilled labor force and new machinery investments by the Japanese. New industrial investment in the region has helped found training programs and created opportunities for young workers, many of whose fathers and grandfathers worked in the same factories, but under very different circumstances. All observers are not optimistic about Northeast China's economic future. Many see the Special Economic Zones of coastal southern China reaping the lion's share of direct foreign investment, leaving little for Shenyang and its region.

CASE STUDY CONNECTIONS

1. Both cities are located on what can be considered cultural and physical frontiers of China.

2. Natural resource exploitation has been an important factor in the industrialization of both cities.

3. Both cities have a distinctive multi-ethnic population, unique in China.

4. Development of Shenyang's rail network was a key to its early growth; Lanzhou did not receive rail service until the mid-20th century.

5. Historically, Lanzhou has served as a stopover point in an extended trade network, whereas Shenyang has served as a regional center.

6. Both cities have been repeatedly occupied by foreigners who have left distinct impressions on the landscape and culture.

7. Both cities are located in harsh environments which place restrictions on the agricultural productivity of the land.

8. Educational levels of the population of Shenyang are much higher than that of Lanzhou.

9. Shenyang's industrial development is largely derived from influences outside of China, whereas Lanzhou's are mostly from within China.

TEST YOUR UNDERSTANDING

Questions

1. Describe the factors involved in Shenyang's industrial development. Describe the factors in its decline.

2. How has Shenyang's relative location in China affected its industrialization both positively and negatively.

3. Lanzhou is often referred to as a cultural crossroads. What factors have contributed to this realization?

Map Exercises

The maps provided at the beginning of this unit can be used to complete this exercise. You may find extra photocopies of these maps helpful. Also, refer to the maps of Chapter 10, "China: The Last Empire?" from de Blij and Muller, 1994, *Geography: Realms, Regions, and Concepts: 7th Edition*, and any standard atlas to assist in this exercise.

1. Place the following on the map provided:

Liaodong Gulf	Fushun	Lanzhou
Northeast China Plain	Liao River	Qinghai
Huang (Yellow) River	Yellow Sea	South Korea
Da Hinggan Mountains	Shenyang	North Korea
Northeast China Mountains	Anshan	Inner Mongolia
Liaodong Peninsula	The Great Wall	Shaanxi Province

2. Map Northeast China's and Gansu Province's natural resources on the map provided. On the same map draw in the transportation infrastructure of both provinces. What does this say about the region's economy? Given the present transportation infrastructure, what avenues are available for international development?

CONSIDER THE DISCIPLINE

The Geography of Resource Conservation

Increasingly in the modern world, **resource management** concerns both **development** and **conservation**. Program 22 addresses both historical and contemporary examples of development in China.

Shenyang's economic base developed around the urban and regional supply of **natural resources**. That is, the physical properties of the natural environment that are considered useful and necessary for satisfying human wants and needs. In Shenyang's case, these physical properties consisted of coal, iron, zinc and other minerals found in the surrounding mountains. Exploiting these resources led to the **industrialization** of the region.

For Shenyang, the process of industrialization has occurred with the development of the natural resource base in the early 20th century. This process was spurred on by the growth of the manufacturing industry in Shenyang. These industries are located near the resources necessary for production in order to minimize transportation costs, a classic example of the economic geography of **regional development.** The location of the steel industry and heavy equipment manufacturing plants in Shenyang attests to this fact. Much of the capital investment that helped drive the industrialization process came from sources outside the country, namely Japan and Russia. When the Communist Party gained control of China in 1949, foreign investment was reduced. Shortly after, the Central Economic Planning Committee also cut economic investments in the Shenyang region, leading to a decline in economic growth for the province.

Critical to competitive industrial growth is the constant **reinvestment** in infrastructure and machinery. Shenyang has recently started reinvesting in these components of its economy, and the economy is once again starting to grow. But the existing infrastructure, educated population and natural resource wealth will not return it to the economic force that it once was without continued financing. The Economic Planning Committee has taken two steps to promote Shenyang's continued resurgence. First, declaring the region an **Open Coastal Area** has encouraged the return of foreign investors, now from Taiwan and Korea, as well as Japan. Second, there are increasing efforts to funnel national funds into Shenyang to promote its growth. Nonetheless, there is debate about whether these measures are enough to keep Shenyang competitive in the world market.

To further develop your understanding of this field of regional geography, refer to the Focus on a Systematic Field in Chapter 10, "China: The Last Empire?" from de Blij and Muller, 1994, *Geography: Realms, Regions, and Concepts: 7th Edition.*

CRITICAL VIEWING

Based on the Case Studies in Video Program 22: "Life in China's Frontier Cities" from *The Power of Place: World Regional Geography,* and the reading in Chapter 4, "The Pacific Rim of Austrasia," from de Blij and Muller, 1994, *Geography: Realms, Regions, and Concepts: 7th Edition,* develop essays answering the following questions:

1. Compare trade routes of three of the cases studies in this unit. Include one Case Study from this program. What comparative advantages exist for each location? How does location enhance and limit trade routes?

2. Define the term "Frontier Zone" and discuss the concept in reference to Gansu Province. Why is the future of Gansu Province and Lanzhou likely to remain regional rather than international in scope?

3. Using the geographic concept of relative location, discuss where you would place a city in Liaoning Province. What factors influence your decision? What role does physical geography play in locating cities? What other factors influence place locations?

■ PROGRAM 23

China's Metropolitan Heartland

LESSONS IN GEOGRAPHY

At the end of Program 23, you should be able to:

1. Discover how rivers and ports can influence the urbanization process.
2. Learn of the development of Shanghai as a major Chinese city.
3. Observe how government planning has affected Shanghai's growth.
4. Examine Shanghai's role in China's national economy.
5. See the effects of industrialization in Sijia, an agricultural village in the Nanjing-Jiangning area.
6. Investigate how non-state, rural township enterprises are changing China.
7. Determine the geographical features that influence the location of rural industrialization.

OVERVIEW

This program examines a prominant city in the **Chang Jiang River** valley, China's heartland, and the vast hinterland it serves. The two Case Studies examine the change that has occured in recent years. *Nanjing-Jiangning: Rural Industry* profiles Sijia, a small village that has seen the rapid growth of a blue jeans garment factory. This **rural township enterprise** has given villagers year-round employment with steady wages and an increasing standard of living.

Shanghai, China's most populous city, is experiencing unprecedented growth since the central government has given its approval and support to the **open city** policy. In *Shanghai: Awakening the Giant,* we see massive construction projects beginning in old Shanghai and throughout this **Open Coastal Area**, especially in the Pudong New Area. This second approach to **economic development,** encouraged by the central and provincial governments, can be compared to the traditional non-state run rural industry in Sijia. The opening of China's coastal regions to **foreign trade** has been the catalyst for the city's economic expansion.

STUDY RESOURCES

1. Video Program 23: "City and Country in China's Heartland" from *The Power of Place: World Regional Geography*.
2. Unit 11 Map, "China and Its Sphere" from Latz and Gilbert, 1996, *The Power of Place: World Regional Geography Study Guide*.
3. Chapter 10, "China: The Last Empire?" from de Blij and Muller, 1994, *Geography: Realms, Regions, and Concepts: 7th Edition*.

PREVIEWING QUESTIONS

1. What relative location factors might contribute to Shanghai's world-class shipping facilities?

2. Why is Shanghai, situated at the mouth of the Chang Jiang River, important to China's economy?

3. What changes could be expected in a rural community undergoing industrialization?

4. How do you think village industrialization will affect the agricultural production of Sijia?

KEY THEMES

- **Chang Jiang's Influence**
- **Urban Shanghai Profile**
- **The Renewed Development of Shanghai**
- **The New Shanghai**
- **Jiangning Township**
- **Rural Enterprises**
- **Village Industrialization**

DISCUSSION OF CASE STUDY THEMES

• Chang Jiang's Influence

The Chang Jiang, or Yangtze, River flows over 3,660 miles (6,000 kilometers) from the highlands of Tibet to the East China Sea, and over 350 million people live in its drainage basin. Shanghai sits at the mouth of this river. The river is an integral part of China's transportation network; it is navigable by ocean-going vessels for over 670 miles (1,100 km) inland to places such as Wuhan and Sijia. Smaller ships can travel another 975 miles (1,600 km) to the industrial city of Chongqing. Together with its tributaries, the Chang Jiang offers over 18,300 miles of navigable waterways that allow for large flows of consumer goods, industrial products and passengers to the hinterland of Shanghai and much of Southeast China. In turn, materials and products flow downstream to Shanghai.

• Urban Shanghai Profile

Shanghai is China's largest city. With a population of over 13 million permanent residents and 3 million itinerant workers, it is also one the largest urban areas in the world. After the Opium Wars between China and the European powers during the 19th century, foreign traders from Britain, France, Russia, the United States and Japan set up operations in Shanghai, turning the city into a booming port and industrial center. Much of Shanghai's unique European/Asian architecture is located along the Bund, a row of tall financial and trade buildings facing the Huangpu, a tributary of Asia's longest river, the Chang Jiang. Today this region is a 21 mile-long port, a mix of oil terminals, container terminals, and bulk loading facilities.

• The Renewed Development of Shanghai

One key to the industrial growth of China's main seaport for international trade is the Pudong New Area, located across the Huangpu east of Shanghai. This tributary of the Chang Jiang was not spanned until the early 1990s. Two bridges now carry vehicles across the river to this Open Coastal Area for which plans include factory sites, a financial center, free-trade zones, residential areas, a new deep water port on the Chang Jiang, a communications center, and an international airport. West of the river, in Old Shanghai, construction is also rapidly occurring. Old buildings are coming down to make way for a new ring freeway and a subway is also being added.

• The New Shanghai

With the rapid change that is occurring in Shanghai, a new city is developing. As the city's infrastructure improves, especially that for transportation, economic opportunity will follow. The Pudong New Area offers many opportunities for further growth. Agricultural fields are giving way to new projects such as high rise apartments, factories and port facilities. This advancement means displacement for residents of older neighborhoods, also a part of redevelopment. Because of its strategic geographic location at the mouth of the Chang Jiang, Shanghai will continue to grow as one of the world's great cities and East Asian economic centers.

• Jiangning Township

Jiangning is one of five counties in the Nanjing Prefecture of the Jiangsu Province. Sijia is one of 24 villages and townships within the county. A small village 19 miles (30 kilometers) south of Nanjing, Sijia produces both agricultural products and blue jeans. The blue jeans garment factory, a rural industrial enterprise or township enterprise, is the result of a long-standing village activity allowing businesses to make profits commesurate with their efforts. Such activity is helping to fuel China's rapid economic growth. In Jiangning County alone there are more than 2,400 rural township enterprises amongst a population of 40,000.

• Rural Enterprises

Most rural enterprises in Jiangning County produce textiles, chemicals or machinery. These factories now employ almost 20 percent of the county. In the village of Sijia, the blue jeans garment factory employs 450 people — over 50 percent of the population — and has had to hire at least 20 percent of its workforce from the neighboring province of Anhui. These migrant workers put in long hours for low pay without the family unit benefits enjoyed by Sijia's residents. Under the government's new policy of freer markets, businesses can make profits commensurate with their efforts. This has helped bring foreign investment to rural enterprises. Sijia's factory attracted it's first group of foreign investors in 1990.

• Village Industrialization

With the success of the blue jeans garment factory, the people of Sijia have been able to obtain many of the comforts afforded by a higher living standard. A new school has been built, many people have built new homes, and many extended families have appliances, washing machines, color televisions, and even VCRs. The factory has helped to keep the rate of migration from Sijia to China's cities lower, since workers can secure a steady income and improve their material wealth within the village.

keep people in small villages instead of migrating to cities

The growth of township enterprises has brought a number of changes to the Chinese countryside. For example, in Sijia, there are few poor people in a country that is still largely poor. The factory has given many, especially women, the chance to earn a steady income. In farming households, having three, four, or even seven family members working in the factory dramatically raises income. This effects the food these families can afford, the clothes they wear and the homes in which they live. More efficient farming methods mean fewer hands are needed in the fields. The factory allows people who would otherweise have to migrate to the city to stay in the countryside and make a good living.

CASE STUDY CONNECTIONS

1. Ethnic homogeneity spans the Chang Jiang River valley. There are few peoples other than Han Chinese in Sijia or Shanghai.

2. China's government now has policies that encourage large-scale projects, as seen in the Open Coastal Area of Shanghai, and that reinforce more traditional small-scale projects, for example, the rural township enterprise in Sijia.

3. Transportation networks are vital to both Sijia's and Shanghai's continued prosperity.

4. Infrastructure development is an important factor in Shanghai's renewed economic might.

5. Both places rely on the Chang Jiang River — Sijia for transportation of trade goods to Nanjing, and Shanghai for access to the 350 million people in the Chang Jiang hinterland.

6. International trade is vital to both places.

7. Access to adequate labor supplies supports both the city and the village.

8. Scales of industrialization and growth are different, yet each location is dependent on both.

9. Both the Open Coastal Area and the rural township enterprise seek foreign direct investment, and both live under its effects.

TEST YOUR UNDERSTANDING

Questions

1. What role has the Chinese government played in Shanghai's recent economic resurgence?

2. Why does the shipping industry play such an important role in Shanghai's economy?

3. How have rural township enterprises affected the standard of living for the people of Sijia?

4. How can rural township enterprises help reduce the rate of rural-to-urban migration in parts of China?

Map Exercises

The maps provided at the beginning of this unit can be used to complete this exercise. You may find extra copies of these maps helpful. Also, refer to the maps of Chapter 10, "China: The Last Empire?" from de Blij and Muller, 1994, *Geography: Realms, Regions, and Concepts: 7th Edition*, and any standard atlas to assist in this exercise.

1. Locate the following:

Chang Jiang River	Huangpu River	Anhui
The Bund	Pudong New Area	Jiangsu
East China Sea	Wuhan	Sijia
Nanjing	The Grand Canal	Suzhou
Huai River	Shanghai	Yichang
Jiangning	Three Gorges Dam Site (New China Dam)	

2. Outline the region of China that is navigable from the mouth of the Chang Jiang. What advantages does this navigable water system give the region?

CONSIDER THE DISCIPLINE

The Geography of Resource Conservation

In addition to being able to feed itself, China can provide many of the raw materials it needs for development. Something as intangible as relative location can also constitute a resource. As we enter the twenty-first century, China's development and world influence will be increasingly dependent upon its resource management.

FACTORS IN URBAN GROWTH

Shanghai's location on the mouth of the Chang Jiang River has allowed the city to dominate Southeast China's economy. The city's **relative location**, along with its **physical geographic features**, has influenced its growth. The Chang Jiang River is the main shipping route to and from much of central China. The large volume of traffic, both goods and people, moving on this waterway is vital to China's economy. Shanghai is able to take advantage of its location by focusing on the shipping industry and container transport. Because of the extensive water transportation network in Southeast China, Shanghai is able to interact with the interior heartland. In addition, Shanghai's access to the ocean facilitates trade with overseas destinations. Because the Huangpu River is wide and deep, it can accomodate many deep-water port facilities, allowing both river and ocean-faring vessels to dock within the heart of the city. These facilities play a vital role in the economy of this government-designated **Open Coastal Area.**

The importance of **infrastructure development** is also apparent in Shanghai. The newly built bridges and tunnels to the Pudong New Area have sparked rapid economic growth. Government involvement, through the financing major projects, has helped to encourage such development.

VILLAGE INDUSTRIALIZATION

China's traditional **rural township enterprises** expand the economic base of many of the small villages in Southeast and South China. Such economic activity allows businesses to keep profits commensurate with their efforts to expand operations and raise wages for their employees. The growth of rural township enterprises throughout China has been one of the key factors in the country's recent period of economic growth. The **effect of industrialization** upon China's many small villages has been dramatic. In the Jiangsu Province, over 90 percent of the population is rural-based. But the township enterprises that are flourishing throughout the region have allowed many of those people to increase their standard of living. Factories provide people with a steady, reliable source of income and higher wages than they would receive for agricultural work. As farming becomes less labor-intensive, the factories give those people once tied to agricultural duties other economic opportunities within their villages. This decreases the rate of **rural-to-urban migration** that many other countries have had to contend with during the industrialization process.

Not all villages have benefited equally from rural township enterprises. **Relative location** is as important in developing rural enterprises as it is in explaining Shanghai's prominence in China's economy. Requirements such as easy access to transportation, adequate labor supply, and proximity to markets are important considerations when locating a new factory.

To further develop your understanding of this field of regional geography, refer to the Focus on a Systematic Field in Chapter 10, "China: The Last Empire?" from de Blij and Muller, 1994, *Geography: Realms, Regions, and Concepts: 7th Edition.*

CRITICAL VIEWING

Based on the Case Studies in Video Program 23: "China's Metropolitan Heartland," from *The Power of Place: World Regional Geography,* and the reading in Chapter 10, "China: The Last Frontier?" from de Blij and Muller, 1994, *Geography: Realms, Regions, and Concepts: 7th Edition,* develop essays answering the following questions:

1. Define the term extraterritoriality. What countries were granted extraterritorial rights in Shanghai? How were these rights obtained? What effects did these rights have on the urban and cultural geography of the city? How will the rights of new foreign-direct investors translate into political influence?

2. Compare the village of Sijia to Dikhatpura, India (Video Program 21). Why has Sijia been able to develop an industrial base while Dikhatpura has not? What role has the government played in each country? How does access to reliable transportation affect each village?

▪ PROGRAM 24

The Booming Maritime Edge

LESSONS IN GEOGRAPHY

At the end of Program 24, you should be able to:

1. Explain the flexibility of the Global Production System.
2. Examine the influence that central government regulation has on the social and economic changes occurring in South China.
3. Learn how Guangdong, a province with three Special Economic Zones, has spearheaded China's drive towards modernization.
4. Identify Taiwan's attempts to control urbanization.
5. Review the development of the Hsinchu Science-Based Industrial Park as featured in Taiwan's urban planning and economic development policies.
6. Observe the importance of research and development in creating new economic and social opportunities for Taiwan.

OVERVIEW

By examining two locations on China's maritime edge, this program explores the recent economic blossoming of South China.

Guangdong: The Booming Maritime Edge profiles how **Special Economic Zones** (SEZs) in this province have been integrated into the **Global Production System** of numerous multinational corporations since Guangdong's first SEZ was established in 1984. Most of the factories in the SEZ profiled are privately owned, a rare occurence in Communist China. Most of the factories have received financial backing from foreign investors, many from Taiwan, and work on a contract basis with foreign trade partners. As **communication technology** has improved, global production is more feasible, and the various functions of a company can be completed at numerous locations. Nike, the American athletic shoe company, has developed a manufacturing and shipping outpost in Guangdong, which provides the backdrop for exploring the flexibility of the Global Production System.

The Republic of China is explored in *Taiwan: Avoiding the Crush*. Taiwan, one of Asia's **economic tigers**, has experienced rapid economic growth and **urbanization** in the past 25 years. As a result, the largest habitable area of the island, **Taipei Basin,** has become congested and polluted, leading many to wonder where future growth in Taiwan will occur. One way in which the government has planned for Taiwan's economic future is through the development of a new **research and development** park, the Hsinchu Science-Based Industrial Park. The effects of such a park on this **newly industrializing economy** will propel Taiwan's already advanced export strategies into the 21st century.

STUDY RESOURCES

1. Video Program 24: "The Booming Maritime Edge," from *The Power of Place: World Regional Geography*.

2. Unit 11 Map, "China and Its Sphere" from Latz and Gilbert, 1996, *The Power of Place: World Regional Geography Study Guide*.

3. Chapter 10, "China: The Last Empire?" from de Blij and Muller, 1994, *Geography: Realms, Regions, and Concepts: 7th Edition*.

PREVIEWING QUESTIONS

1. Where is Guangdong located in the realm of China?

2. What is a Global Production System?

3. How might urban planning help to avoid problems such as congestion and air pollution?

4. How can research and development provide economic security for Taiwan's future?

KEY THEMES

- **Special Economic Zones in Guangdong Province**
- **The Global Production System**
- **Importance of Infrastructure**
- **Communication Technology**
- **Taipei Basin**
- **Hsinchu Science-Based Industrial Park**
- **National Development for the 21st Century**

DISCUSSION OF CASE STUDY THEMES

• Special Economic Zones in Guangdong

International trade has become one of the most important sectors of Guangdong's economy. Coastal and international shipping routes link over 100 large and small ports in the province, of which Guangzhou and Shantou have national significance. The importance of these ports was emphasized when the government declared Shenzhen, Shantou, and Zhuhai Special Economic Zones as a part of economic reforms implemented in 1979 to realize Deng Xiao Ping's Four Modernizations. These SEZs are designed to take advantage of foreign capital by allowing more economic freedoms, such as duty-free trade, tax incentives, relaxation on foreign investment restrictions, and permitting profits made in China to be expatriated to the investors' home countries. Guangdong Province, with its overseas links through emigration, trade history, and proximity to the international financial connections of Hong Kong, has helped these SEZs take advantage of their status. The result has been tremendous economic growth throughout these zones in Guangdong, growth which at present drives the rest of China.

growing regional disparities vast contrast

• The Global Production System

The Global Production System has been developed to maximize efficiency in the production process. The system is highly organized, with numerous activities proceeding simultaneously in a variety of locations. As orders are being taken in one location, components of the final product are being made in a variety of other locations. The parts are then shipped to a manufacturing facility that combines the parts to make a finished product, which in turn is sent to a port for shipment to a distribution center, and finally to a retail location.

• Importance of Infrastructure

A well-developed infrastructure is crucial to Guangdong's continued economic growth. Without modern facilities including ports, highways, buildings and telecommunications networks, companies cannot compete in the world economy. The delays and resulting expenses caused by poor transportation systems threaten the cost advantages that lure industry to these locations. Poor communications networks can slow information transfer, causing problems in other factories in the global production system.

• Communications Technology

Communications technology is central to the Global Production System. This technology has the ability to link geographically diverse operations through the use of fiber-optics and faxes which monitor computerized inventories. In the Nike example, bar-codes read into a computer in Hong Kong can be downloaded through the telecommunication lines to a central computer in Beaverton, Oregon, where decisions regarding production and marketing are made and products are tracked.

• Taipei Basin

Taiwan's capital is the city of Taipei, located in northwestern Taiwan. The city sits in one of the coastal basins of western Taiwan, ringed by mountains. The rapid economic development of Taiwan has produced a relatively wealthy population, but has also produced some unwanted side effects. Rising affluence has meant more motor vehicles. The surrounding mountains trap emissions from these vehicles, creating a heavy blanket of air pollution. A second problem created by the mountains is that of space. Some are concerned that further expansion in the Basin will strangle the city's economic growth.

• Hsinchu Science-Based Industrial Park

Hsinchu Science-Based Industrial Park (HSIP), 28 miles (45 kilometers) southwest of Taipei, was developed by the government to give high-tech businesses a place to build new factories away from the congestion of Taipei. HSIP is the home to more than 200 high-tech companies. The Park, situated to take advantage of the proximity of Taipei, the international airport, and major seaports, is part of the Taiwanese government's plan to lure more high-tech industries to Taiwan. Local universities, the government-funded Industrial Technology Research Institute, a new infrastructure, and the high standard of living associated with the Park all help draw new businesses to the Park and attract highly trained workers to the island.

• National Development for the 21st Century

The government invested in HSIP in order to entice new businesses to the Republic of China. As this newly industrializing economy grew, the cost of doing business in export-oriented light industries (such as textiles) increased. Taiwan's competitors took advantage of this by producing goods at a cheaper rate. Facing increasing competition, increasing dependence on raw materials from abroad and increasing salaries, Taiwan looked to new strategies to reduce its vulnerability to external economic conditions. In order to stay ahead, in 1980 the country embarked on a program to develop more capital and technology-intensive industries. Domestic investments such as HSIP are one approach toward advancing the economy into a new stage.

CASE STUDY CONNECTIONS

1. Both locations are far from China's cultural hearth associated with the area of Beijing.

2. Both have Hong Kong as a major trading partner.

3. Taiwan is financing operations within SEZs and Open Coastal Areas in China and is one of many countries investing in Guangdong Province.

4. Government planning has been crucial to the continued development of both locations.

5. Many emigrants have left Guangdong for other areas of Pacific Asia; Taiwanese emigrants have moved mainly to North America.

6. Taiwan is trying to encourage immigrants with high-technology knowledge to return and use that knowledge in the Hsinchu Science-Based Industrial Park.

8. Foreign investment in the Guangdong Province is often by Chinese who have emigrated from that region.

9. Both Guangdong and Taiwan are linked to the world economy through the Global Production System and international trade.

10. Both locations play a vital role in the continued growth of economies in East Asia, Southeast Asia, and the world.

GEOGRAPHERS AT WORK

Richard Barff is an economic geographer at Dartmouth College whose work focuses on the global forces of trade and investment driving regional development in China and in other rapidly growing areas in the world.

Guangdong Province, the object of Barff's research, is China's most rapidly growing region, with gross industrial output growth rates averaging in the double digits from 1979 to the present. Contained in it is the epicenter of such development, known as the Pearl River (Zhu Jiang) Delta, immediately adjacent to Hong Kong. Guangdong and its delta region which includes Guangzhou are being touted by some academics as having the potential to lead the province to become the fifth "Asian dragon."

The key to the heady growth of Guangdong is its ability to attract foreign investors who seek access to its cheap labor supply and favorable and flexible regulatory environment. The province's proximity to Hong Kong is also important in its own right because Hong Kong is a steady source of capital and access to international markets. Truly global forces drive the growth of Guangdong, but its future will be determined by the future political stability of the Peoples Republic of China as well.

TEST YOUR UNDERSTANDING
Questions

1. How do communication and information transfers allow the Global Production System to operate?
2. Why would a company like Nike want to contract shoe manufacturing to a separate business instead of producing the shoes itself?
3. What benefit does the Taiwanese Government see in creating a high-technology industrial park?
4. What benefits are there for a Taiwanese entrepreneur living in the United States who is starting a business in the Hsinchu Science-Based Industrial Park?

Map Exercise

The maps provided at the beginning of this unit can be used to complete this exercise. You may find extra copies of these maps helpful. Also, refer to the maps of Chapter 10, "China: The Last Empire?" from de Blij and Muller, 1994, *Geography: Realms, Regions, and Concepts: 7th Edition*, and any standard atlas to assist in this exercise.

1. Place the following names:

Taiwan	Guangzhou	Hong Kong
Guangdong Province	Xi Jiang	Chung-yang Mountains
Taipei	Nan Mountains	Pei River
Kao-hsiung	Taiwan Strait	Bashi Channel
Chan-chiang	South China Sea	Hsi River
Shantou	Zhuhai	Shenzhen
Zhu Jiang (Pearl River)		

CONSIDER THE DISCIPLINE

Our Geography of Resource Conservation discussion continues with an eye to how **relative location** and **abundant labor** function as resources in the world economy. Guangdong and Taiwan are premier examples of the world economy at work. Both have economies that are heavily dependent on **international trade** for continued growth and future prosperity. Their regional positions on international trade routes depend upon their distance, accessibility, and connectivity to their global markets.

Taiwan has created a newly developing economy by taking advantage of the **globalization of production**. Starting in the early 1950s, Taiwan began reorienting its industries toward light manufacturing of products such as textiles. Through careful management of profits and reinvestment, the economy has slowly improved its standard of living; it is now one of the highest in Asia.

Guangdong has taken a different path toward the globalization of its economy. China's Communist government was slow to allow the country to participate in the world economy. However, in recent years, under Deng Xiao Ping's Four Modernizations policy, the creation of Special Economic Zones has allowed Guangdong to participate more freely in the world economy. Since, then growth of South China has been dependent upon **foreign investment** and international trade.

Globalization of the local economy has allowed Guangdong to exercise its comparative advantage in **low-wage labor**. The rapid growth that Guangdong has experienced since the late 1980s, estimated to be above nine percent per year, has helped to improve the region's living standard by increasing wages and providing new jobs.

To further develop your understanding of this field of regional geography, refer to the Focus on a Systematic Field in Chapter 10, "China: The Last Empire?" from de Blij and Muller, 1994, *Geography: Realms, Regions, and Concepts: 7th Edition.*

CRITICAL VIEWING

Based on the Case Studies in Program 24: "The Booming Maritime Edge," from *The Power of Place: World Regional Geography,* and the reading in Chapter 10, "China: The Last Empire?" from de Blij and Muller, 1994, *Geography: Realms, Regions, and Concepts: 7th Edition,* develop essays answering the following questions:

1. While both Sijia (in Program 23) and Guangdong have recently been able to prosper, they have achieved success through very different methods. Compare and contrast the policies of the Chinese government as related to the village of Sijia and the province of Guangdong.

2. Taiwan is making plans to supplement its manufacturing industry as places such as Guangdong begin to lure foreign investors away from Taiwan. Should Guangdong start to think about supplementing its manufacturing industry as well? What stages of growth does Guangdong need to go through in order to catch up to Taiwan?

▪ UNIT 12

Southeast Asia: Between the Giants

SOUTHEAST ASIA:
Between the Giants

| Kilometers | 0 | 350 | 700 |
| Miles | 0 | 200 | 400 |

MYANMAR
(BURMA)

Red R.

Hanoi

Irrawady R.

20°N

Laos

Chao Phraya

Mekong R.

Rangoon

T H A I L A N D

Bangkok

CAMBODIA

Andaman
Sea

Phnom Penh

Gulf of
Thailand

Vietnam

10°N

South

China

Sea

PACIFIC

20°N

OCEAN

THE PHILIPPINES

LUZON

Manila

Philippine

Sea

10°N

MINDANAO

Straits of
Malacca

Kuala
Lumpur

S U M A T R A
(SUMATERA)

Malaysia

Singapore

BRUNEI

B O R N E O

Celebes
Sea

EQUATOR

EQUATOR

CELEBES
(SULAWESI)

INDIAN

Jakarta

Java Sea

Banda Sea

OCEAN

J A V A
(JAWA)

Indonesia

TIMOR

95°E

110°E

125°E

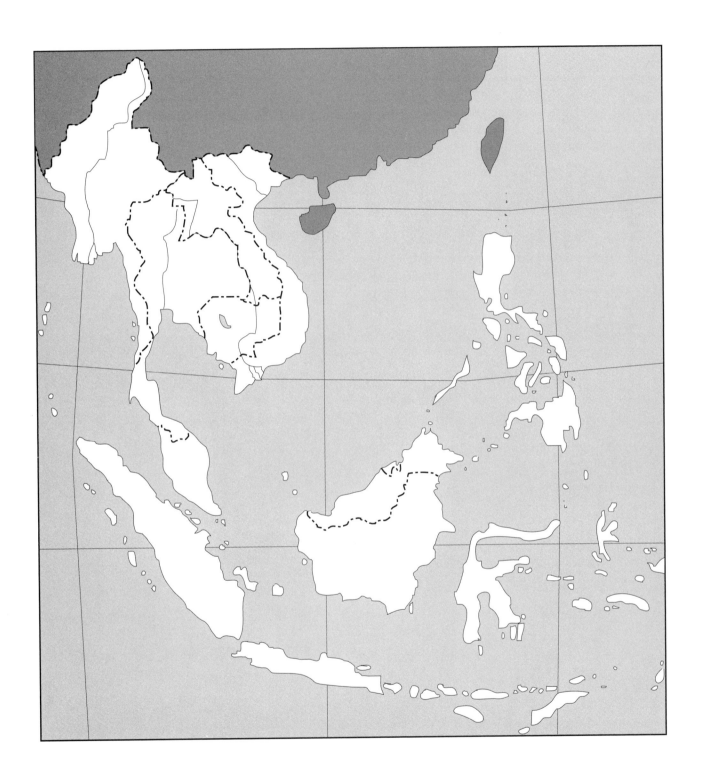

■ PROGRAM 25

Mainland Southeast Asia

LESSONS IN GEOGRAPHY

At the end of Program 25, you should be able to:

1. Identify the factors that made Laos an isolated country both within Southeast Asia and in the world.

2. Examine the effects that the Friendship Bridge, linking Laos to Thailand across the Mekong River, may have on Laos' economic development.

3. Observe the effects of isolation on health and education in Laos.

4. Describe how Laos is attempting to link itself to the economic development occurring throughout the Southeast Asian realm by building infrastructure such as bridges, roads, and hydroelectric power-generating facilities.

5. Discuss Vietnam's recent agrarian reforms which have given farmers more incentive to produce large rice crops and have helped catapult Vietnam into the third largest rice exporter in the world.

6. Follow the transportation routes of the rice crop on roads and canals from the fields of the Mekong Delta to the milling plants in towns and cities.

7. Understand the importance of the Mekong Delta's water resources in rice farming and the irrigation practices which can improve crop yields.

8. Examine the role of the Vietnamese government in the agrarian process.

OVERVIEW

In this program we visit two countries on the mainland of Southeast Asia: Laos and Vietnam. Both countries are attempting to open their borders to the world after years of isolation due to civil war, political unrest, and physical geographic barriers.

In *Laos: Isolated Heart* we see years of **isolation** resulting from its physical geographic barriers — the **Mekong River** and the mountains which cover 70 percent of this traditionally poor country. Political unrest and military conflict in the **Southeast Asian Realm** during the latter half of the 20th century have also served to limit access to Laos. **Landlocked** by five other countries, Laos has no sea access and must rely on ground and air travel to interact with other countries. The results of this isolation can be seen in Laos' **agrarian economy** where, with little economic development, most of the population lives in small villages. Even the capital, Vientiane (Viangchan), has a population of fewer than 150,000 people. With the recent opening of the Friendship Bridge over the 1,100-mile Mekong River, overland access to Thailand is available for the first time. Other road linkages are planned to connect Laos to China and Vietnamese ports. Developers hope to exploit the country's vast hydroelectric power-generating potential to serve markets in Thailand and eventually within Laos itself.

In *Vietnam: Fertile Dreams,* the agrarian sector of the economy — rice production in the **Mekong Delta** — has been transformed. In the past 15 years, the Vietnamese government has engineered a remarkable **agrarian reform.** Improved management of the water from the sediment-rich Mekong River for **irrigation,** combined with a move away from **collective farming** to **contract farming,** have made Vietnam the third largest rice **exporter** in the world. Farmers contract land for a period of 20 years, producing up to three crops per year and are provided with short-term loans to cover their overhead by the government-run Vietnam Bank for Agriculture. In addition to irrigation, the water of the Mekong Delta also provides convenient **transportation,** allowing crops to be brought to mills and markets quickly and inexpensively. As agriculture has found renewed economic vitality in the Mekong Delta, the rest of southern Vietnam has also prospered. Foreign firms have been reopening branch offices in Ho Chi Minh City, known as Saigon to most of its estimated 3 million residents. The **open door policy**, implemented in 1986, has helped to revive the manufacturing and industrial sectors. But the key to the Vietnamese economy remains the rice production process, dependent on a complex interaction of water resource management, transportation, finance, and organization at both the household and village levels.

STUDY RESOURCES

1. Video Program 25: "Mainland Southeast Asia" from *The Power of Place: World Regional Geography.*
2. Unit 12 Map, "Southeast Asia: Between the Giants" from Latz and Gilbert, 1996, *The Power of Place: World Regional Geography Study Guide.*
3. Chapter 12, "Southeast Asia: Between the Giants" from de Blij and Muller, 1994, *Geography: Realms, Regions, and Concepts: 7th Edition.*

PREVIEWING QUESTIONS

1. Where is the Mekong River located in relation to Laos and Vietnam?
2. How do climate and water dictate techniques of rice production?
3. When did colonial rule end in the former Indochina?
4. How has military conflict shaped both Laos and Vietnam?

KEY THEMES
- **Isolation**
- **The Mekong River**
- **Poverty in a Rural Country**
- **Creating a Country Linked to the World**
- **Water: Lifeblood of Rice Production**
- **Farming Reform**
- **Bringing the Rice to Market**
- **Water Resource Management**

cultural heart - Annam

3 important cities — Hue, Hanoi, Saigon - economic capital

DISCUSSION OF CASE STUDY THEMES

Handwritten note in left margin: 2nd lowest population density in SE Asia

• Isolation

The mountain ranges that surround Laos to the north, east, and west, the country's landlocked nature, and limited access across the Mekong River have served to physically isolate Laos from other countries. The combination of Laos' physical characteristics with a long period of civil war and political strife both within the country and among many of its neighbors has caused the economy to stagnate in the decades following World War II. New government policies aimed at creating an open-market system and a push toward improved transportation links with other countries, for example the Friendship Bridge crossing the Mekong River at Vientiane, are expected to help revive the Laotian economy.

• The Mekong River

The Mekong River is one of the geographical keys to Laos. The river helps to define the country's borders, provides soil and water for wet-field rice production, and is a major transportation route for Laotian people and products. The river not only insulates the country from the outside world, but acts as a barrier to progress by limiting access to Laos due to a lack of bridges. In addition, catastrophic floods regularly occur over the river's wide floodplain.

  The hydroelectric plant executive, Somphavan Inthavong, claims that the Mekong may be the link that will tie Laos to the outside world. Power production on the river far exceeds domestic consumption, creating a surplus for neighboring markets. Mr. Inthavong, who has selected 60 promising dam sites, expects the country to substantially increase its social welfare through economic development of the Mekong's hydropower.

• Poverty in a Rural Country

Laos is one of the poorest counties in the world. There is little industrial activity within Laos, and that activity is confined to a few cities, such as Vientiane. Most Laotians live in rural, agricultural villages and work as farmers. Besides religion, there are few opportunities for education, and few Laotians are literate. Health care is also affected. There are few hospitals in the countryside, and low immunization coverage means that Laos suffers from one of the highest infant mortality rates in the world.

• Creating a Country Linked to the World

The Laotian government is pursuing a course that will help to link the country to the world. The building of the Friendship Bridge is the first step in providing land linkages to other countries. Plans are being formulated to build roads through the mountains in the east and north, providing links to cities in Vietnam and China. As Laos becomes connected to other countries, it remains to be seen how it will overcome the challenges associated with industrialization and economic development.

• Water: Lifeblood of Rice Production

The Mekong River provides Vietnamese farmers with many of the necessities for wet-field rice farming. The river's delta has been formed by the huge quantity of sediment carried by the river. As the river slows and reaches the South China Sea, the sediment is dropped, leaving rich, fertile soil. The frequent flooding of the rice fields renews the soil and allows for numerous harvests each year. But it is the water itself that is crucial to wet-field rice farming, and insuring that the appropriate amount of water reaches fields at the right time is vital to production of this aquatic plant.

• Farming Reforms

Vietnam is moving away from the collective farming arrangements set up after the Indochina (Vietnam) War. The new contract system allows farmers to lease land for 20 year intervals and to keep any profits resulting from their efforts. Community cooperation remains pivotal during the labor-intensive planting and harvest times.

Government organizations such as the Bank of Agriculture help to provide independent farmers with enough funds to plant their crops. The government also provides capital to build milling facilities and to improve transportation systems which move the grain.

• Bringing the Rice to Market

The numerous natural channels and artifical canals of the Mekong Delta along with roads built by the French and later, the Americans, provide farmers with inexpensive transport routes to the major rice markets of southern Vietnam. The Mekong River also grants ocean-going vessels access to port facilities, where rice can be loaded for destinations around the world.

• Water Resource Management

Vital to the three harvests produced by many of the Mekong Delta's rice fields is the appropriate use of river water. Because the timing, quantity and quality of water application are vital to the growth of rice crops, farmers, with government support, have created water cooperatives which share pumping and piping facilities and improve irrigation methods.

CASE STUDY CONNECTIONS

1. Laos has a population of 4.7 million. Neighboring Vietnam has a population of 72.3 million.

2. The Mekong River flows through China, Cambodia and Laos, and its Delta lies in southern Vietnam.

3. Agriculture is the primary activity in each country. Subsistence farming is commonly practiced in Laos, while contract farming is practiced in Vietnam.

4. Both have entered a new era of seeking foreign markets and investment, but Vietnam is further along the path to becoming a newly industrializing economy.

5. Laos endures physical isolation. Vietnam is emerging from global isolation.

6. Water plays a key role in the economic development of each country.

TEST YOUR UNDERSTANDING

Questions

1. Laos hopes the Friendship Bridge will help the country's economic development. Why is this one bridge important to Laos' economic future?

2. How has the isolation of Laos affected community services within the country?

3. What needs does the Mekong River fulfill for the Vietnamese rice farmers who populate the Mekong Delta?

4. How has the government of Vietnam been able to help farmers increase rice production in the Mekong Delta?

Map Exercises

The maps provided at the beginning of this unit can be used to complete this exercise. You may find extra copies of these maps helpful. Also, refer to the maps of Chapter 11, "Souteast Asia: Between the Giants," from de Blij and Muller, 1994, *Geography: Realms, Regions, and Concepts: 7th Edition*, and any standard atlas to assist in this exercise.

1. Place the following names:

Mekong River	Annamese Cordillera	Hanoi
Red River	Vientiane (Viangchan)	Thailand
China	Ho Chi Minh City (Saigon)	Laos
Mekong Delta	Hué	Cambodia
Myanmar (Burma)	South China Sea	Friendship Bridge
Gulf of Thailand	Mount Bia	Vietnam

CONSIDER THE DISCIPLINE

POLITICAL GEOGRAPHY: TERRITORIAL MORPHOLOGY IN MAINLAND SOUTHEAST ASIA

The **elongated** shape of Vietnam influences its cohesion, spatial organization, and political viability. The north-south spatial variation of the **physiography** of Vietnam has influenced political events from a traditionally subordinate relationship to China, more recent colonial divisions and post-independence civil war. As Vietnam moves to join world agricultural and other markets, the coastal orientation of this nation positions the entire country along the dynamic Pacific Rim.

The **insular** nature of Laos is much like that of island nations. Instead of being surrounded by water, this landlocked country is **surrounded** by five other nations. What was previously a disadvantage in the country's development may now be a favorable location for exporting the surplus of hydroelectric power generated by the Mekong River to Laos' industrializing neighbors.

To further develop your understanding of this field of regional geography, refer to the Focus on a Systematic Field in Chapter 11, "Southeast Asia: Between the Giants," from de Blij and Muller, 1994, *Geography: Realms, Regions, and Concepts: 7th Edition*.

CRITICAL VIEWING

Based on the Case Studies in Video Program 25 "Mainland Southeast Asia" from *The Power of Place: World Regional Geography,* and the reading in Chapter 11, "Southeast Asia: Between the Giants," from de Blij and Muller, 1994, *Geography: Realms, Regions, and Concepts: 7th Edition,* develop essays answering the following questions:

1. The Mekong River, the longest in Southeast Asia, originates in the mountains of southwestern China. Compare the lifestyles of the people who live along the river as it flows from its source through China, Myanmar, Laos, Thailand, Cambodia and Vietnam.

2. Many people fled South Vietnam when the Nationalist government fell in 1975. Refugees settled in Thailand, Malaysia, Indonesia, and the United States. As the Vietnamese economy becomes increasingly active in the world, this diaspora may play a vital role in the country's development. Refer to earlier programs to compare this scenario to the role played by expatriates from Vancouver, Singapore, Taiwan and Guangdong in the economies of their homelands.

▪ PROGRAM 26

Maritime Southeast Asia

LESSONS IN GEOGRAPHY

At the end of Program 26, you should be able to:

1. Discuss the origins of Malaysia's multi-ethnic society.
2. Explain how the balance is maintained among Malaysia's ethnic groups.
3. Describe the importance of Malaysia's location to its economy.
4. List the factors at play in Malaysia's rapidly growing and industrializing economy.
5. Examine the growing importance of tourism in the Indonesian economy.
6. Describe how the influx of foreign travelers has changed the landscape of Bali.
7. Compare the living conditions of the population of Jakarta, Indonesia's capital, to those of Bali, Indonesia's major tourist destination.

OVERVIEW

Program 26 focuses on two maritime countries in Southeast Asia, Indonesia and Malaysia. The cultural impact of economic development poses a unique challenge to these countries. While each struggles to become industrialized, the ethnic composition and cultural variety of their people present their own promises and problems.

Indonesia: Tourist Invasion examines one economic opportunity being developed and supported by the government in Bali, an island in the Indonesian archipelago. **Tourism's** benefits to Bali include training and employment for the Balinese people and the attraction of foreign currency to Indonesia through both foreign visitors and property investment. A large resort area, Nusa Dua, has been built to serve as a magnet for tourists from across the globe. The growth of the Nusa Dua complex provides **employment opportunities** for both the local population and other Indonesians while diverting migration from the crowded capital city, Jakarta. But tourist development is not without problems. Kuta, the island's oldest tourist area, displays **overdevelopment**, **crowding**, **cultural dislocation**, and **environmental damage** — the results of tourism gone rampant. With the influx of large numbers of foreigners, Balinese worry about the erosion of traditional culture, religion, and family values.

Multi-cultural Malaysia examines that country at the historical crossroads of maritime trade between China, India, and the Arabian Peninsula and presents the country's **ethnic composition**: 60 percent Malay, 30 percent Chinese, and 10 percent Indian. The village of Rengit, 183 miles (300 km) southeast of the capital city of Kuala Lumpur, provides insight into the relationship among people of different ethnic groups. Many farmers in the region are ethnic Malay who grow oil palms, the main source of income for most of Malaysia's farmers. In Rengit, as in much of

Malaysia, ethnic Chinese often act as commerical brokers, in this case for the local oil palm trade. As more ethnic Chinese have begun to develop business opportunities, their growing economic power and land holdings have changed the structure of the agricultural and urban sectors of Malaysia. The resulting Chinese **economic dominance** led to riots in 1969, pitting Malays against Chinese. To reduce the economic gap among different ethnic groups and to promote social harmony, the government embarked on a policy of *Bumiputera* or "Malays First." The policy begins with education and employment and extends across the entire range of social relations. Rather than allowing economic forces to create disproportionate power for one ethnic group over others, Malaysia's policy of Bumiputera strives to balance **customs**, **lifestyles** and **educational opportunites**, even as preferences for the Malay majority are promoted. The resulting peaceful coexistence of Malaysia's different ethnic and religious groups has contributed to the country's booming economy and remarkable growth over the past 25 years.

STUDY RESOURCES

1. Video Program 26: "Maritime Southeast Asia" from *The Power of Place: World Regional Geography*.

2. Unit 12 Map, "Southeast Asia: Between the Giants" from Latz and Gilbert, 1996, *The Power of Place: World Regional Geography Study Guide*.

3. Chapter 12, "Southeast Asia: Between the Giants" from de Blij and Muller, 1994, *Geography: Realms, Regions, and Concepts: 7th Edition*.

PREVIEWING QUESTIONS

1. Which European colonial powers occupied these two countries? For how long?

2. Where is the equator in relation to this region?

3. How many islands comprise the nation of Indonesia? Where are Java (Jawa) and Bali located in the archipelago?

4. Is Malaysia an island country?

KEY THEMES

- **Tourist Development**
- **Shift from Subsistence Economy**
- **External Influences**
- **Translocation**
- **Multi-ethnic Social Policy in Malaysia**
- **Economic Opportunity for All**

DISCUSSION OF CASE STUDY THEMES

• Tourist Development

The Indonesian government has taken steps to foster development in areas peripheral to the country's major cities. One such step has been state-supported programs to expand the Balinese economy through the tourist industry. State-sponsored programs help to obtain investment capital, which is used to build tourist developments like the Nusa Dua complex. Jobs created from these developments help relieve the population pressures on Jakarta by providing employment close to home for the Balinese people and by luring people away from the population centers of Java.

• Shift from Subsistence Economy

Before the tourist invasion, Bali's primary economic activity was fishing. Now the emphasis of the Balinese economy has shifted from working to meet the needs of the local residents toward meeting the needs of temporary visitors. Seventy percent of the locals now depend on tourism for their income. Low-wage, low-skill jobs predominate in this industry. Yet, since many hotels seek a skilled local management team, training opportunities are provided. A village resident who, in an earlier era, might have learned fishing techniques from the elders, now might learn computer techniques from a global hospitality chain.

• External Influences

Tourist development is not without problems. The influx of foreigners can erode the local culture and lifestyles, as witnessed in the city of Kuta, Bali's oldest tourist area. Street hawkers, glowing neon signs, drugs, and prostitution have transformed this surfer's paradise from the quiet home of farmers and fisherman to what some call tourism gone rampant.

• Translocation

Bali has witnessed a large influx of Javanese. Java, with 120 million residents, is the most populous of Indonesia's 13,000 islands. In fact, the island of Java is among the world's most densely populated areas. The island is also home to Indonesia's power base — the capital city of Jakarta. With nearly 10 million inhabitants, Jakarta seeks to relieve its urban population pressures through a government policy encouraging resettlement of Javanese to other Indonesian islands. This process, called translocation, places those sympathetic to the Jakarta government's interests in distant positions of power throughout this far-flung country.

• Multi-ethnic Social Policy in Malaysia

The government of Malaysia has developed the policy of Bumiputera to help the country's different ethnic groups coexist. The policy is designed to allow the Malays, Chinese, and Indians that constiture Malaysia's population to maintain their own customs and religious practices while coexisting. By implementing this policy across a range of social relations, primarily in the educational system and the workplace, the government has been able to close economic gaps among the country's ethnic groups.

Business owners are required to provide Muslim Malays with opportunities to practice their religion during the workday. In schools, courses in Malay language are compulsory, while additional courses have been designed to foster inter-ethnic understanding. This policy encourages people to maintain their own customs while accepting the customs of others. This has resulted in the creation of more opportunities for ethnic Malays. As these opportunities arise, Malays, who historically have lived in rural areas, are beginning to migrate to towns and cities, where opportunities are available in industrial and manufacturing sectors.

• Economic Opportunity for All

Malaysia is a newly industrializing economy. The rapid growth of the Malaysian economy has helped to decrease ethnic friction. It has afforded virtually the entire population opportunities to increase their standard of living and has fostered ties among ethnic groups through business partnerships and transactions.

CASE STUDY CONNECTIONS

1. Both Indonesia and Malaysia are considered to be newly industrializing economies which are making a transition from resource extraction to manufacturing and industrial production.

2. For centuries, both countries have been connected to distant lands through trade.

3. Both are former colonies of European powers that gained independence in the 1940s.

4. The government of both countries play a vital role in fostering economic activity.

5. While both countries share the island of Borneo with Brunei, Malaysia is connected to the mainland of Asia, while Indonesia is completely insular.

6. Both countries are adjacent to the equator.

7. Agricultural production in Indonesia supplies a variety of crops such as coffee, tea, tobacco, spices, and rice. In Malaysia, the main crops are oil palms and rubber.

8. The majority of the population of both countries is of Muslim faith.

9. There are large populations of other minority groups in both countries, ethnic Chinese and Indians in Malaysia, and religious Hindus and tribal people in Indonesia.

GEOGRAPHERS AT WORK

Nagata Junji is a Japanese geographer at the University of Tokyo studying the population characteristics of Malaysia, focusing on ethnic diversity.

Nagata observes that ethnic diversity is only one aspect of a country's population; others might include age, gender, occupation, etc. What distinguishes study of ethnicity in Malaysia's case, however, is the fact that the economic and population policies of the country are intertwined by government decree through the policy of Bumiputera. Whereas, in many countries, ethnic studies document the protests by ethnic groups, often minorities, against government policy, Malaysia is a country where ethnic issues are being carefully managed in order to create a society that balances the interests of its Malay majority and its smaller, but disproportionately financially more powerful, Chinese minority.

The economic progress of Malaysia depends on the quality of its labor force, but in the geographic research presented here there is an essential cultural component that cannot be ignored.

TEST YOUR UNDERSTANDING

Questions

1. Why does the Indonesian government want to promote tourist developments such as Nua Dusa in Bali?

2. How has the town of Ubud been able to avoid the problems that Kuta has seen with the growth of the tourist industry on Bali?

3. Describe some of the restrictions imposed upon ethnic Chinese through Bumiputera.

4. List the different groups that have controlled Malaysia and the imprints that they left on Malaysian culture.

Map Exercises

The maps provided at the beginning of this unit can be used to complete this exercise. You may find extra copies of these maps helpful. Also, refer to the maps of Chapter 11, "Southeast Asia: Between the Giants," from de Blij and Muller, 1994, *Geography: Realms, Regions, and Concepts: 7th Edition*, and any standard atlas to assist in this exercise.

1. Place the following names:

South China Sea	Malay Peninsula	Sumatra
Strait of Malacca	Borneo	Bali
Kuala Lumpur	Pinang Island	Jakarta
Singapore	Java	New Guinea
Celebes Sea	Kalimantan	Brunei

2. Create a map showing the Southeast Asian countries once controlled by each of the European colonial powers.

CONSIDER THE DISCIPLINE

POLITICAL GEOGRAPHY

To the geographer, a country's existence is predicated on the idea of the state expressed through political control of a **defined territory**. National identification of the state as a particular place with established **boundaries** is always complicated by **physical and cultural geography**. Southeast Asia is a case in point: a region with a host of states newly emerging from long histories of colonial domination. A central problem in these newly formed countries is the integration and control of territories that cover vast peninsular and island areas and are populated by an array of ethnic groups with distinct languages and customs. Where common interests binding the state are strong, **centripetal** forces are at play and the state succeeds; where the ties that bind are weak, **centrifugal** or divisive forces interfere and the state fails. Policies which promote national identity, such as Bumiputera in Malaysia, are an example of a centripetal force encouraging social and economic development.

Consideration of the physical and cultural geographical circumstances in Indonesia and Malaysia helps to define the political prospects for each state as well as underscores their social and economic development policies, focused on tourism and ethnic coexistence, respectively.

Indonesia, a republic in Southeast Asia, is composed of over 13,500 islands. The 3,200 mile-wide country straddles the equator from the Indian Ocean to Papua New Guinea on the island of New Guinea. In the north, the Strait of Malacca, South China Sea, Celebes Sea and Pacific Ocean separate Indonesia from the mainland of Asia. The major Indonesian islands are composed of rugged volcanic mountains covered by dense tropical forests. In the south, these islands include Sumatra, Java, Madura, Bali, Lombok, Sumbawa, Flores, and Timur, while the northern portion of Indonesia is composed of the islands of the northern Moluccas, New Guinea, Celebes and the southern portion of Borneo, called Kalimantan.

Indonesia's recorded history began when Malays first migrated to the islands almost 1,000 years ago. Commercial relations were common with the surrounding countries, including China and India, leading to the cultural influences of Hinduism and Buddhism. Muslim traders made contact with Sumatra during the 13th Century, and by the 16th Century, the Portuguese, Spanish, Dutch, and British were all trading with the islanders. The Dutch East India Company eventually acquired colonial control of most of Indonesia, and the Dutch controlled the islands for some 300 years until the mid-twentieth century. In 1949, the United States of Indonesia was formed with nominal ties to the Netherlands. These ties were dissolved in 1954 over disputes regarding the control of western New Guinea. In 1965, a military coup placed General Suharto in command of the country, and he continues to act as president into 1995.

Malaysia is a country divided. Located in the Southeast Asian Realm, it is split by 400 miles of the South China Sea. The eastern states (approximately the size of Oregon) are located on the island of Borneo which they share with Indonesia and Brunei. Across the sea to the west lies the second half of Malaysia. The country encompasses the southern end of the Malay peninsula from the Thailand border in the north to the city-state of Singapore, a former territory that separated from Malaysia in 1965, in the south. Controlled by a succession of foreign powers from 1511 to until the British relaxed colonial rule in 1948, Malaysia became what is now known as the Federation of Mayalsia in 1966. The country is strategically located on the Strait of Malacca. This Strait, long vital to trade between east and west, contains some of the busiest waters in the world. The multi-ethnic fabric of Malaysia has been woven from these east-west crossroads which have historically attracted immigrants from India, China and Arabia. This country's ethnic mix and its present-day policy of Bumipatera can be traced to these geographic forces.

To further develop your understanding of this field of regional geography, refer to the Focus on a Systematic Field in Chapter 11, "Southeast Asia: Between the Giants," from de Blij and Muller, 1994, *Geography: Realms, Regions, and Concepts: 7th Edition.*

CRITICAL VIEWING

Based on the Case Studies in Video Program 26: "Maritime Southeast Asia" from *The Power of Place: World Regional Geography,* and the reading in Chapter 11, "Southeast Asia: Between the Giants," from de Blij and Muller, 1994, *Geography: Realms, Regions, and Concepts: 7th Edition*, develop essays answering the following questions:

1. Forecast the future of Balinese culture. Will this island's heritage be lost, change or evolve?

2. Will multi-ethnic harmony endure in Malaysia? As Malaysia joins the Pacific Rim, will its approach to cultural accomodation diffuse across international markets?